Environmental Aspects, Applications and Implications of Nanomaterials and Nanotechnology 2015

Topical Conference at the 2015 AIChE Annual Meeting

Salt Lake City, Utah, USA
8-13 November 2015

ISBN: 978-1-5108-1848-4

Printed from e-media with permission by:

Curran Associates, Inc.
57 Morehouse Lane
Red Hook, NY 12571

Printed by Curran Associates, Inc. (2016)

For permission requests, please contact AIChE
at the address below.

AIChE
120 Wall Street, FL 23
New York, NY 10005-4020

Phone: (800) 242-4363
Fax: (203) 775-5177

www.aiche.org

Additional copies of this publication are available from:

Curran Associates, Inc.
57 Morehouse Lane
Red Hook, NY 12571 USA
Phone: 845-758-0400
Fax: 845-758-2634
Email: curran@proceedings.com
Web: www.proceedings.com

TABLE OF CONTENTS

428598 Application of Sub-Microfluidic Devices in Water Filtration Processes to Investigate the Removal of Pathogens

Abstract:

Access to clean water has become a critical issue as the water resources became scarce. There are several filtration techniques used in water treatment facilities such as reverse osmosis, and sand and gravel filters. Slow sand filtration is an economical filtration step used to obtain drinking water in cities. It filters particles and bacteria as water flows through the sand. Studies revealed that bacteria could alter their shape and fit into structures that are narrower than their sizes. The effect of flow on bacterial morphology in pores is not well documented. Here, we are applying nanobiotechnology to investigate the bacterial behavior in narrow structures as they are exposed to fluid flow. Miniaturized systems are great tools to mimic pores, and observe bacterial cells at a single cell level. This study will provide a pressure profile for pathogenic bacteria to improve the filtration processes. We designed sub-microfluidic devices that have 950 nm tall constrictions with widths ranging from 1.5-3 μm. The constrictions connect the inlet channel, where the pressure is applied, to an outlet channel, which is open to atmosphere.

The microfluidic devices were fabricated in the clean room. Constrictions were patterned on a silicon wafer with electron beam lithography, and chromium metal was sputtered to determine the height of the constrictions. Inlet and outlet microchannels were aligned onto the constrictions with photolithography. The completed master wafer was used as a mold to obtain poly-dimethylsiloxane (PDMS) devices, which were then permanently bonded to glass cover slips. PDMS is a polymer that is widely used in microfluidics field to study cells. Here, we studied two microorganisms: *Escherichia coli* (*E. coli*) and *Pseudomonas aeruginosa* (*P. aeruginosa*). *E. coli* is an indicator of contamination when detected in water. The bacteria were initially grown overnight in lysogeny broth at 37 °C. The cultures were diluted to avoid accumulation of bacterial cells at the entrance of the constrictions. The cells were flowed through the constrictions as the pressure was applied at increasing increments by using Fluigent microfluidics flow control system, which was connected to a house airline. The images were taken after each pressure application.

Our results show that filtration of species varies with the applied pressure. When the cells are exposed to the same applied pressure, the physiology of bacteria becomes a critical factor to determine their behavior in narrow structures. *E. coli*, which is larger in diameter and less motile than *P. aeruginosa*, entered the constrictions at higher pressure values. As the constriction size became narrower than the diameter of both bacteria, the pressure applied significantly increased. Further analysis is critical to determine optimum parameters for separating pathogens to improve the filtration processes.

430313 Electrocatalytic Inactivation of E. coli in Point of Use Drinking Water Applications Using TiO2 Nanotubes

Abstract:

A flow reactor device initiating electroporation was developed for bacterial disinfection in lab-synthesized and natural waters. Complete inactivation of bacterial contaminated waters with flow rates of 50 mL/min was observed. The device consisted of immobilized titanium dioxide nanotube arrays, which applied a bias of up to 5 V without noticeable water splitting. Water chemistry, applied voltage, and bacterial concentration were all found to impact disinfection rates. Complete inactivation of *E. coli* W3110 (~8 x 10^3 CFU/mL) occurred in 15 seconds in the device with an applied voltage of 3 V in a 100 ppm NaCl solution. Inactivation rates varied widely as a function of ionic concentrations. Disinfection in natural water was inhibited by the presence of inorganic ions and other constituents which are commonly found in natural water. A higher voltage of 6 V was required to reach 100% efficiency because of this. An applied bias to the nanotube flow through disinfection chamber shows promise as an effective bacterial disinfection device for *E. coli* inactivation without the need for solar irradiation.

STABILITY OF CARBON DOTS IN NATURAL WATERS: IMPLICATIONS FOR SENSING AND THE ENVIRONMENT

María M Fidalgo, Jingjing Dai, Austin Zambrana and Chloe Rees, Department of Civil and Environmental Engineering, University of Missouri, Columbia, MO, US.

Introduction

Interest for both fundamental research and applications has surged dramatically carbon nanodots (C-dots)[1-3]. Thus, it is inevitable such materials will find their way to natural water sources where, unlike the more well-studied fullerenes and carbon nanotubes, their environmental impact and fate is mostly unknown.

C-dots constitute a fascinating class of recently discovered nanocarbons that comprise discrete, quasi-spherical nanoparticles with sizes below 10 nm[1]. Because C-dots display size and excitation wavelength-dependent photoluminescence behavior, can be derived from inexpensive and abundant materials, and have presumably lower environmental impact, they are attractive alternatives to metal-based quantum dots (such as CdS and CdTe) especially for bio applications and aqueous-phase sensing. Early studies have shown C-dots are biocompatible[4] and less cytotoxic than Cd-based quantum dots[5] and may even inhibit cancer cells[6]. The increase in activity for these emerging carbon nanomaterilas make their appearance in natural waters inescapable.

Due to their large surface area and activity, adsorption is expected to play an important role in the fate and transport of C-dots [7]. It has been shown that adsorption may alter the surface characteristics of nanoparticles, so that they become dispersed and mobile in aqueous media, or aggregate and settle rapidly [8]. Organic macromolecules, such as natural organic matter, polysaccharides and proteins, are ubiquitous in natural waters and therefore readily available to interact with nanomaterials [9].

The objective of this work is to investigate the stability of C-dots, an emerging carbon nanomaterial, in natural waters (i.e. variable salinity) and the effect of the transformations they undergo under such conditions on the fluorescence properties.

Methods and Materials

Synthesis of C-dots:

Carbon Dots were synthesized following a modification of a previously published method [10]. A 70% w/w glycerol solution (Sigma, >99.5%) was added to a 20% w/w solution of NaH_2PO_4 (Fisher Scientific). The solution was heated for 20 minutes at 900W power in a domestic microwave. The product was allowed to cool to room temperature and the C-dots resuspended by addition of water and continuous stirring until a homogenous colored suspension was obtained.

Amine capped C-dots (AC-dots) were also fabricated: 2.10 g of a 10 mmol solution of citric acid (Sigma-Aldrich ACS reagent, >=99.5%) was added into 10 mL of Polyethylenimine, ethylenediamine branched (Sigma-Adrich). The reaction was heated from room temperature to 170 °C in silicone oil bath (Fisher Scientific) for 2h. Then the reaction flask was naturally cooled to room temperature and deionized water was added. The resulting solution was dialyzed against water

(Spectrumlabs, MWCO 1.0 kDa) for 2 days for purification, replacing the water was every 6 hours. Finally, AC-dots were freeze-dried to yield a yellow solid.

Size and surface charge measurements

Size measurements were performed by Dynamic Light Scattering (DLS) using a Zetasizer Nano ZS, Malvern, UK. Laser Doppler velocimetry was applied to characterize the electrophoretic mobility (EPM) of the nanomaterials in the various electrolyte solutions using a Malvern Zetasizer Nano ZS. Measured EPMs were converted to *zeta*-potential using the Smoluchowski equation [11]

$$U = \varepsilon\zeta / \mu$$

where U is the electrophoretic mobility, ε is the dielectric constant of the solution, μ is its viscosity, and ζ is the zeta potential. Disposable folded capillary cells were employed. The samples were taken from the freshly prepared suspensions, but measured after at least one hour, when aggregation equilibrium was assumed.

Fluorescence measurements

Fluorescence readings were conducted in a Hitachi F-4500 Fluorescence Spectrophotometer. For C-dots emission scans, an excitation wavelength of 350nm was used, corresponding to the peak from the excitation scan. For excitation scans, the emission wavelength used was 450nm for C-dots, and 460nm for AC-dots. These values correspond to the emission peaks. A slit width of 5nm was used for all measurements. Scans were conducted in a wavelength range of 300nm to 700nm, at a speed of 1200 nm/minute.

Results and Discussions

Two synthesis routes for C-dots were used and investigated. In the first approach, a solution of glycerol is thermally treated in the presence of a catalyst to yield the nanoparticles (C-dots). Surface functionalization is commonly performed after C-dots synthesis, either to give special functionality to the particle itself or to provide sites for reaction with other chemicals. Amino groups are frequently used for this last strategy. The second synthesis method used in this work yielded amino-capped carbon dots (AC-dots).

The transport behavior of the nanomaterials in the environment is expected to be controlled partially by their chemical composition but mostly by their surface chemistry. Although chemically similar with respect to their core composition, these two nanoparticles have the potential to react very differently to changes in natural water chemistries, and the effect on their stability and inherent properties (fluorescence) may also be modified.

Fresh waters (rivers, lake) are characterized by low ionic strength (1mM -5 mM) while seawater ionic strength can be as high as 0.5 M. Groundwater has intermediate, variable levels. The effect of increasing ionic strength on the size and surface potential, characterized by the zeta potential, of the carbon dots was investigated and shown in Figure 1 and figure 2 respectively; NaCl and $CaCl_2$ were used as electrolytes. The dots were resuspended in the electrolyte solutions from their "as-prepared" dry forms, resulting in a pH of circumneutral for the C-dots and a pH of 9 for AC-dots.

Increasing ionic strength is expected to have a destabilization effect in charged particles due to the compression of ionic double layer that turns electrostatic repulsion forces shorter range and allows for the net interaction energy to be attractive.

That trend was weakly observed for both materials in NaCl solutions; AC-dots became very unstable at ionic strengths higher than 50 mM and a reliable value for their size could not be obtained by DLS due to the presence of large aggregates (> 5 microns) and fast settling. The effect was even less clear in the divalent cation solutions. One possible explanation for the apparent indifference with respect to ionic strength may be their very small size, which may result in other mechanisms than surface charge (e.g. diffusion) being the ones controlling the aggregation phenomenon in the range of ionic strengths investigated.

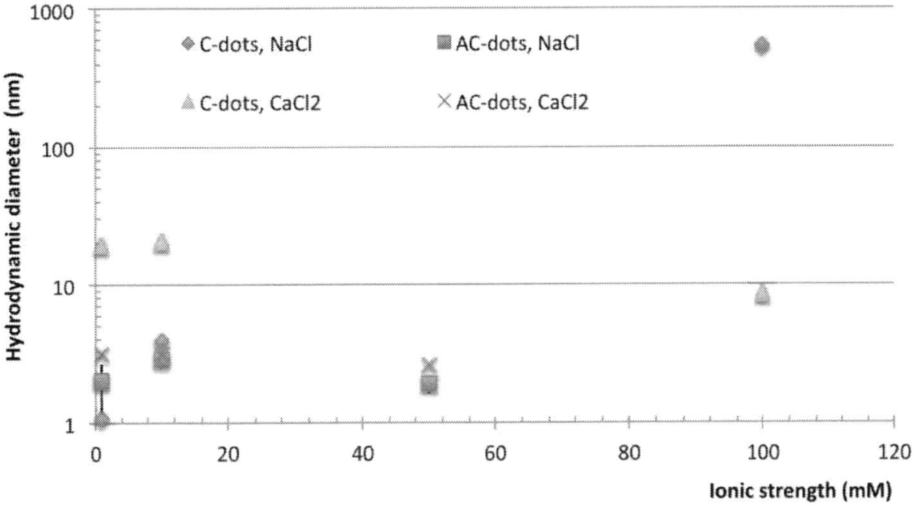

Figure 1. Hydrodynamic diameter of C-dots at various ionic strength concentrations; C-dots: glycerol derived carbon dots; AC-dots: amino capped carbon dots.

However, the effect of ionic strength in the zeta potential of the particles was important. As expected, C-dots showed a negatively charged surface wile amine carbon dots were positively charged; the absolute value of the zeta potential in deionized water was significant in both cases and resulted in a strong stabilization effect. The zeta potential of C-dots followed the expected trend of decreasing absolute values with increasing salt concentration. The effect was even more pronounced in the presence of Ca^{2+}. Monovalent cations (Na^{+}) are considered to indifferent with respect to the surface, while some degree of interaction due to specific adsorption may occur in the presence of divalent cations. In particular, calcium is well known for promoting aggregation of dissolved natural organic matter (NOM). Due to chemical similarities of NOM with the C-dots, we hypothesized that calcium may also play a role in the destabilization of carbon nanomaterials. Calcium ions seemed to adsorb onto the negatively charged C-dots, giving a lower surface charge, in absolute values. While the decrease in zeta potential for the Na^{+} solutions could be entirely attributed to charge shielding due to double layer compression, an additional effect of specific adsorption was clearly present in the Ca^{+2} suspensions. Interestingly, the opposite was observed for the positively charged AC-dots. Increasing concentrations of NaCl electrolyte produced a significant decrease in surface charge, the presence of divalent cations resulted in a reduced decrease in charge. This observation hints to some degree of adsorption of the Ca^{2+} to the positively charge surface. In this way, there is an additional contribution to charge in the particle surface that compensates, at least partially, the expected zeta potential decrease due to higher electrolyte concentrations.

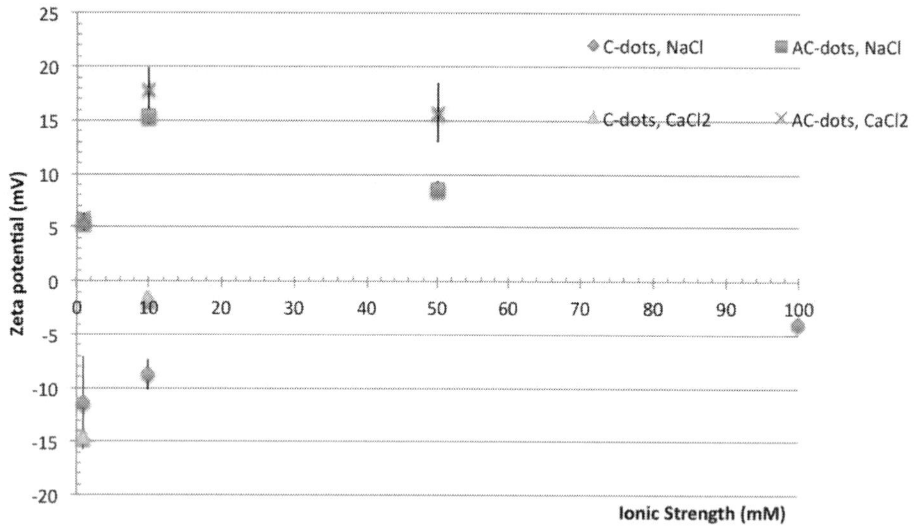

Figure 2. Effect of ionic strength on the surface charge of C-dots; C-dots: glycerol derived carbon dots; AC-dots: amino capped carbon dots.

Fluorescent nanoparticles are especially interesting in sensing applications in environmental science as well as medical imaging, due to the high emission intensity and stability. Quantum dots have been extensively investigated but their toxicity in aqueous systems makes them fairly unattractive. Carbon dots are a benign alternative fluorescent particle with a huge potential in both fields. In order to be useful, their properties need to remain stable over the range of conditions normally encountered in natural waters as well as physiological mediums.

We investigated the fluorescence emission intensity of both nanomaterials in three levels of ionic strength, given by two electrolytes: NaCl and $CaCl_2$. The results are presented in Figure 3. Since aggregations negatively affects the emission, we focused our work on those ionic strength conditions in which we did not observed significant particle size growth in our DLS determinations. Fluorescence emission intensity increased with ionic strength for both C-dots and AC-dots in the presence of an indifferent electrolyte. However, different effects were observed for the suspensions prepared in $CaCl_2$. Lower emission intensities were observed for C-dots with increasing concentration of Ca^{+2} in comparison with the same ionic strength level given by the monovalent ion. However, the presence of calcium resulted in a negligible effect on the AC-dots. The specific adsorption of Ca^{+2} was much more noticeable on the negatively charged C-dots that in the AC-dots; there was, nevertheless some adsorption as indicated by the differences in zeta potential detected for the latter, but it was not strong enough to have an influence on the fluorescence properties of the carbon dots.

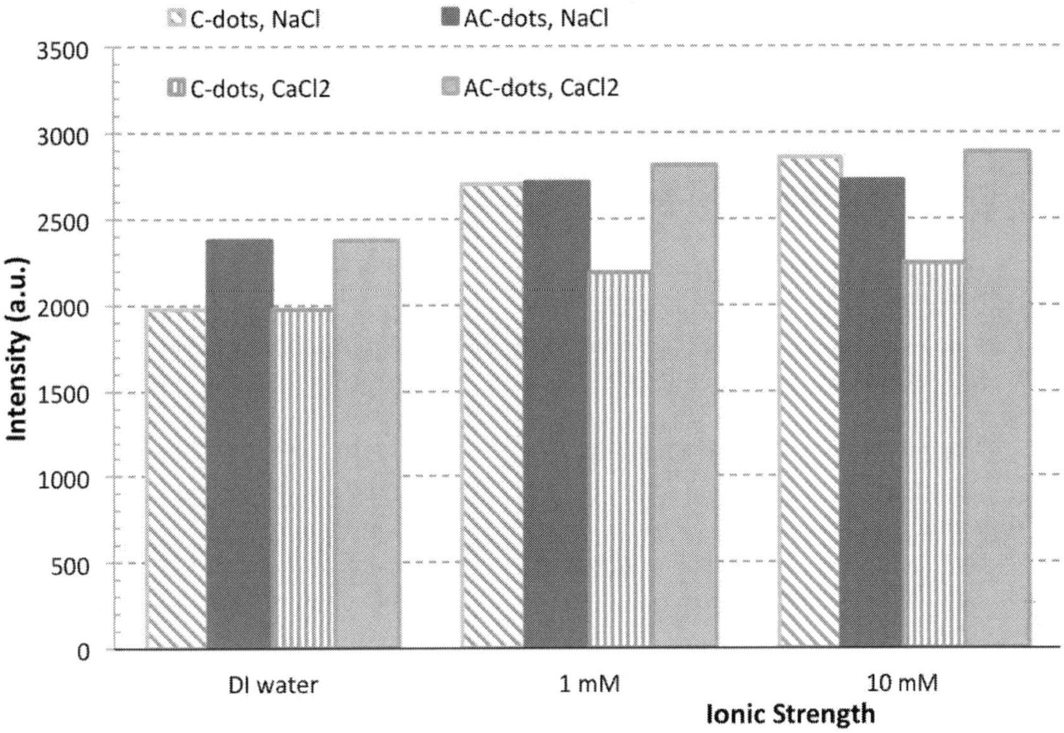

Figure 3. Fluorescence intensity peak for carbon dots at various levels of ionic strength. C-dots: glycerol derived carbon dots; AC-dots: amino capped carbon dots.

Conclusions

The carbon nanomaterials investigated, C-dots and AC-dots, resulted in stable suspensions up to an ionic strength level of 50mM, both in the presence of monovalent and divalent cations. Calcium ions can adsorb specifically to both materials, regardless of the fact that the AC-dots are positively charged: however, the adsorption and its effect on stability and fluorescence was much more noticeable for the negatively charged C-dots.

The increase in electrolyte concentration produced a decrease in absolute value of the zeta potential of the particles; in the presence of calcium ions, this charge screening effect is accentuated for C-dots and counteracted for AC-dots. Given the stabilization effect of surface charge, we can expect C-dots to become unstable and aggregate in calcium rich waters, while AC-dots would exhibit increased stability and therefore their higher tendency to stay in the water column and be transported with the flow.

The emission intensity exhibited a small increase over the ionic strength levels investigated in the presence of a non-adsorbing electrolyte. However, in the calcium rich suspensions, a decrease of fluorescence intensity was evident for the C-dots only. It is unclear if the reason for the observed variations is related to some surface interactions with the electrolyte or if the somewhat larger particle size observed for C-dots in $CaCl_2$ suspensions (i.e. moderately aggregated state) may have produced the lower emission observed.

References

1. Baker, S.N. and G.A. Baker, *Luminescent carbon nanodots: Emergent nanolights.* Angewandte Chemie - International Edition, 2010. **49**(38): p. 6726-6744.
2. Li, D. and R.B. Kaner, *Materials science: Graphene-based materials.* Science, 2008. **320**(5880): p. 1170-1171.
3. Rao, C.N.R., A.K. Sood, K.S. Subrahmanyam, and A. Govindaraj, *Graphene: The new two-dimensional nanomaterial.* Angewandte Chemie - International Edition, 2009. **48**(42): p. 7752-7777.
4. Qu, S., X. Wang, Q. Lu, X. Liu, and L. Wang, *A biocompatible fluorescent ink based on water-soluble luminescent carbon nanodots.* Angewandte Chemie - International Edition, 2012. **51**(49): p. 12215-12218.
5. Song, Y., D. Feng, W. Shi, X. Li, and H. Ma, *Parallel comparative studies on the toxic effects of unmodified CdTe quantum dots, gold nanoparticles, and carbon nanodots on live cells as well as green gram sprouts.* Talanta, 2013. **116**: p. 237-244.
6. Hsu, P.C., P.C. Chen, C.M. Ou, H.Y. Chang, and H.T. Chang, *Extremely high inhibition activity of photoluminescent carbon nanodots toward cancer cells.* Journal of Materials Chemistry B, 2013. **1**(13): p. 1774-1781.
7. Chowdhury, I., M.C. Duch, N.D. Mansukhani, M.C. Hersam, and D. Bouchard, *Colloidal properties and stability of graphene oxide nanomaterials in the aquatic environment.* Environmental Science and Technology, 2013. **47**(12): p. 6288-6296.
8. Lanphere, J.D., C.J. Luth, and S.L. Walker, *Effects of saolution chemistry on the transport of graphene oxide in saturated porous media.* Environmental Science and Technology, 2013. **47**(9): p. 4255-4261.
9. Saleh, N.B., L.D. Pfefferle, and M. Elimelech, *Influence of biomacromolecules and humic acid on the aggregation kinetics of single-walled carbon nanotubes.* Environmental Science and Technology, 2010. **44**(7): p. 2412-2418.
10. Wang, X., K. Qu, B. Xu, J. Ren, and X. Qu, *Microwave assisted one-step green synthesis of cell-permeable multicolor photoluminescent carbon dots without surface passivation reagents.* Journal of Materials Chemistry, 2011. **21**(8): p. 2445-2450.
11. Elimelech, M., J. Gregory, X. Jia, and R.A. Williams, *Particle Deposition and Aggregation.* Colloid and Surface Engineering: Applications in the process industries. Controlled Particle, Droplet and Bubble Formation, ed. R. Williams. 1995, Linacre House, Jordan Hill, Oxford, UK: Butterworth-Heinemann, Ltd.

EXPERIMENTAL STUDIES USING ULTRAVIOLET-VISIBLE SPECTROSCOPY ON SYNTHESIZED NANOPARTICLES

Engr. Dr. M. N. Idris* and **A. Abdullahi**
Energy & Process Refining Technology Group
Department of Chemical Engineering
University of Maiduguri
Borno State, Nigeria
Tel: +234 (0) 705 877 1096
Email: idrismn@unimaid.edu.ng, muhinu@hotmail.co.uk

ABSTRACT

In this study, silver nanoparticles were synthesized by two main chemical reduction methods. Silver nitrate was adopted as the main precursor, and reduced by sodium borohydride and trisodium citrate to produce particles of different size regimes. UV-Visible spectrophotometer was used to monitor the formation of the nanoparticles. The synthesis of nanoparticles via sodium borohydride was carried out at both ambient temperature and ice-bath temperature to compare the effect of temperature on the quality of final synthesis. The stability of the silver nanoparticles dispersed in aqueous medium was investigated with respect to time.
A yellow hue was given off by the silver nanoparticles sols that, a typical surface Plasmon resonance absorption provide maxima of silver nanoparticles at 370-450 nm spectra. The sizes of the nanoparticles were found to be dependent on the reduction routes. The sodium borohydride and trisodium reduction routes resulted in silver nanoparticles with a diameter of around 12 nm and 35 nm, respectively. However, sodium borohydride sols were found to be more concentrated and had lower polydispersity. In the borohydride synthesis, lower synthesis was found to be helpful in formation, purity and stability of the nanoparticles colloids. Also, an increase in the volume ratios of aqueous precursor solutions was found to result in nanoparticles colloids that were more concentrated, less distributed and had higher colloidal stability.

Keywords: *Colloidal, Characterization UV-visible, Synthesis & Polydispersity*

Correspondence about this article*: Please for all correspondence on this article, kindly contact the principal author as stated above. Thank you.*

1.0 INTRODUCTION

A particle is defined as a small object that behaves as a whole unit in terms of its transport and properties [1]. Particles are further classified according to size in terms of diameter [1]. Course particles cover a range between 10,000 and 2,500 nm. Fine particles are sized between 2,500 and 100 nanometers. Ultrafine particles or nanoparticles are sized between 1 and 100 nm. The basis of the 100 nm limit for nanoparticles is the fact that novel properties that differentiate them from the bulk material typically develop at a critical length scale of under 100 nm [2, 3].

The quest to fabricate, characterize and understand small noble metal molecules is among the outstanding challenges of contemporary material science, as such nanoparticles bridge the atomic and the bulk worlds [4]. A bulk material should have constant physical properties regardless of its size, but at the nano-scale size, dependent properties are often observed. Thus, the properties of materials change as their size approaches the nanoscale and as the percentage of atoms at the surface of the material becomes significant. The interesting properties of nanoparticles are therefore, largely due to the larger surface area, which dominates the contributions made by the small bulk of the materials. [5].
Nanoparticles are small enough to confine their electrons and produce quantum effects and can absorb much light than material in thin films of continuous sheets. Those fascinating properties make them Physico-chemically superior to both atomic and bulk materials of the same element [6]. For instance, copper which is opaque at macro scale becomes completely transparent to visible light at the nanoscale [7]; stable materials like aluminum turn combustible [8]; Gold, which is rarely soluble becomes more soluble [9].
In this research, samples of silver nanoparticles are synthesized and their formation is studied using UV-Visible spectrophotometer.

1.1 Problem Statement

Metal nanoparticles, due to their special properties and also small dimensions, find important applications in optical, magnetic, thermal, sensoric devices, catalysis, etc. Many metal nanoparticles are under active research because they possess interesting physical properties differing considerably from that of the bulk phase [10]. It comes from the small sizes and high surface/volume ratio. The most common method employed for the synthesis of metal nanoparticles is the reduction of metal ions in solution [11, 12]. However, the control of size and shape at the nanometer level is a real problem, as the mechanism of size/shape control is still left largely unresolved.
In this work, an attempt has been made to analyze formation of silver nanoparticles from silver salt as well as after storing for variable time in atmospheric condition. The effect of time after synthesis on characteristics of these particles is investigated Using UV/Visible spectrophotometer. Further, the effect of synthesis temperature and using an alternate reducing agent is analysed.

1.2 Ultraviolet-Visible Spectroscopy

Spectroscopy is a technique used to quantify the light that is absorbed and scattered by a sample (a quantity known as the extinction, which is defined as the sum of absorbed and scattered light). In its simplest form, a sample is placed between a light source and a photo detector, and the intensity of a beam of light is measured before and after passing through the sample. These measurements are compared at each wavelength within UV-Visible range (300-700 nm) to quantify the sample's wavelength dependent extinction spectrum. The data is typically plotted as extinction as a function of wavelength. Each spectrum is background corrected using a "blank" – a cuvette filled with only the dispersing medium – to guarantee that spectral features from the solvent are not included in the sample extinction spectrum [13].
Nanoparticles have optical properties that are sensitive to size, shape, concentration, agglomeration state, and refractive index near the nanoparticle surface, which makes UV/Vis spectroscopy a valuable tool for identifying, characterizing, and studying these materials. Nanoparticles made from certain metals, such as gold and silver, strongly interact with specific wavelengths of light and the unique optical properties of these materials is the foundation for the field of plasmonics [13].

1.3 Aims and Objectives

The aim of this project is to study the formation of silver nanoparticles using Ultraviolet-visible spectrophotometer.
The objectives are:

- To synthesize samples of silver nanoparticles.

- To study the effect of varying precursor volumes on the resulting nanoparticles.
- To study the effect of synthesis temperature on the resulting nanoparticles.
- To study the effect of an alternate reducing agent on the resulting nanoparticles.
- To study the effect of storage for variable times in the atmosphere.

1.4 Scope of the Research

The scope of this research is synthesis of colloidal silver nanoparticles by reduction method (using two reducing agents) at two different temperatures; and characterization of all samples synthesized using UV-Visible spectrophotometer.

2.0 BRIEF BACKGROUND ON NANOPARTICLES

Until the Middle Ages, the reputation of soluble gold was based mostly on its fabulous curative powers of various diseases, for example, heart and venereal diseases, dysentery, epilepsy, and tumors; it was also used in the diagnosis of syphilis. The history of the nanoparticles from ancient times to the Middle Ages has been summarized by Daniel and Astruc [15]. The first book on colloidal gold was published in 1618 by the philosopher and medical doctor Francisci Antonii. This book includes considerable information on the formation of colloidal gold sols and their medical uses, including successful practical cases. The book noted that soluble gold appeared around the fifth or fourth century B.C. in Egypt and China. On the other hand, industrial manufacturing of stained glass with colloidal particles was established by Kunckel in the seventeenth century (1676). He also published a book whose Chapter 7 was concerned with 'drinkable gold that contains metallic gold in a neutral, slightly pink solution that exerts curative properties for several diseases' [16]. He concluded that gold must be present in aqueous gold solutions to a degree of contamination such that it is not visible to the human eye. A colorant in glasses, that is, the 'Purple of Cassius', was a colloid resulting from the presence of gold particles and tin dioxide and was highly popular in the seventeenth century. A complete treatise on colloidal gold was published in 1718 by Helcher [17].

In 1857, in a well-known publication, Michael Faraday [18] reported the formation of deep red solutions of colloidal gold by reduction of an aqueous solution of chloroaurate ($AuCl_4^-$) by phosphorus in CS_2 (a two-phase system). He also investigated the optical properties of thin films prepared from dried colloidal solutions and observed reversible color changes of the films upon mechanical compression (from bluish-purple to green). Since that pioneering work, thousands of scientific papers have been published on the synthesis, modification, properties, and assembly of metal nanoparticles, using a wide variety of solvents and other substrates [19].

Nanotechnology is easily evident in various old churches. A well-known application of early nanotechnology is the ruby red color that was used for stained glass windows during the Middle Ages. Beautiful examples of these applications can be found in glass windows of many Gothic European cathedrals, among which the León Cathedral (Spain) represents one of these unique masterpieces, located on the medieval French pilgrimage path to Santiago de Compostela (Spain); its impressive 2000 m^2 colored windows offer a unique view that certainly warrants a visit. Of course, the medieval artisans were unaware that they were using nanotechnology. They just knew that a particular process produced a beautiful effect. For example, the stained glass of a wonderful rose can be seen at the world heritage Cathédrale Notre-Dame de Chartres in France. Later, chemistry clarified the reasons behind the generation of the color. These vivid colors were controlled by the size and the form (or shape) of the nanoparticles of gold and silver. The relation between particles and their associated colors has been discussed recently by Jin and coworkers [20]. In an article of 22 February 2005, the New York Times summarized the relationship between the color of stained glass and the size/shape of the nanoparticles [21].

After several decades, the ingredients present in the stained glass (colored glass) of various churches were clarified subsequent to the development of analytical instruments. People without professional expertise in nanotechnology are also increasingly contributing to the technology. This continues as we speak, with various nanostructures being proposed and discovered, and their applications described [19].

2.1 Properties of Nanoparticles

2.1.1 Uniformity

The chemical processing and synthesis of high-performance technological components for the private, industrial, and military sectors requires the use of high-purity ceramics, polymers, glass-ceramics, and composite materials. In condensed bodies formed from fine powders, the irregular particle sizes and shapes in a typical powder often lead to non-uniform packing morphologies that result in packing density variations in the powder compact [22].

Uncontrolled agglomeration of powders due to attractive van der Waals forces can also give rise to in microstructural inhomogeneities [23]. Some pores and other structural defects associated with density variations have been shown to play a detrimental role by growing and thus limiting end-point densities. Differential stresses arising from inhomogeneous densification have also been shown to result in the propagation of internal cracks, thus becoming the strength-controlling flaws [24]. It would, therefore, appear desirable to process a material in such a way that it is physically uniform with regard to the distribution of components and porosity, rather than using particle size distributions that will maximize the green density. The containment of a uniformly dispersed assembly of strongly interacting particles in suspension requires total control over inter-particle forces. Monodisperse nanoparticles and colloids provide this potential. Inert gas evaporation and inert gas deposition are free many of these defects due to the distillation (cf. purification) nature of the process and having enough time to form single crystal particles, however even their non-aggregated deposits have lognormal size distribution, which is typical with nanoparticles [25].

2.1.2 Large Specific Surface Area

Nanoparticles possess strong surface area effects (e.g. reactivity, high energy of surface area, adsorption, higher solubility, lower melting point, etc). For bulk materials larger than one micrometer (or micron), the percentage of atoms at the surface is insignificant in relation to the number of atoms in the bulk of the material, while the properties of materials change as their size approaches the nanoscale and as the percentage of atoms at the surface of a material becomes significant. Example, the high surface area to volume ratio of nanoparticles provides a tremendous driving force for diffusion, especially at elevated temperatures. Sintering can take place at lower temperatures, over shorter time scales than for larger particles [25].

2.1.3 Quantum Effects

Nanoparticles often possess unexpected optical properties as they are small enough to confine their electrons and produce quantum effects [26]. For example gold nanoparticles appear deep-red to black in solution. Nanoparticles of yellow gold and grey silicon are red in color; Gold nanoparticles melt at much lower temperatures (~300 °C for 2.5 nm size) than the gold slabs (1064 °C); Absorption of solar radiation is much higher in materials composed of nanoparticles than it is in thin films of continuous sheets of material. In both solar PV and solar thermal applications, controlling the size, shape, and material of the particles, it is possible to control solar absorption [27, 28].

Other size-dependent property changes include quantum confinement in semiconductor particles, surface plasmon resonance in some metal particles and superparamagnetism in magnetic materials. What would appear ironic is that the changes in physical properties are not always desirable. Ferromagnetic materials smaller than 10 nm can switch their magnetisation direction using room temperature thermal energy, thus making them unsuitable for memory storage [29].

2.2 Types and Origin of Nanoparticles

There are a number of ways, based on different aspects, that NPs can be classified into groups. Considering their origin, they can be divided into natural and anthropogenic (man-made) which, in turn, can be subdivided into accidental and engineered or manufactured NPs or they can be further separated, based on their chemical composition, into organic (carbon containing) and inorganic [30]. As the name clearly implies, natural NPs are a group of nanosized materials which are generated by the action of natural processes and have been present in the natural environment for a very long time since the origin of the planet Earth. In addition to natural NPs, earlier human activities like mining, agriculture, forest burning and constructions have tremendously increased the concentration of accidental nanoparticles into the environment especially into the atmosphere. Despite the fact that NPs have been present in the environment for millions of years as natural and later as accidental NPs, it was not until quite recently that their interest in both pure and applied sciences have been extensively exploited due to the discovery of their promising, novel and applicable properties which make them attractive for a vast number of products in a wide range of scientific researches and for many sectors of the modern technology.

2.2.1 Natural Nanoparticles

Natural NPs are omnipresent and they are distributed throughout the atmosphere, oceans, soil systems, terrestrial water systems (groundwater and surface water), and in and/or on most living organisms both at the micro and macro levels and have been present in the environment forever [31]. They have been formed mainly by the effect of naturally occurring physical, chemical and biological processes such as hydrolysis, erosion, weathering, volcanic eruption, sea spray, photochemical reactions, growth nuclei in super-saturated fluids, plants roots on rocks, minerals and microorganisms. Forest fires, as volcanic eruptions, for instance, can spread ash and smoke over thousands of miles and lead to an increase in particulate matters, including NPs, exceeding ambient air quality standards [32].

2.2.2 Accidental Nanoparticles

Human activities have been releasing different forms of nanoparticulate matter for millennia as by-products of some activities which were essential for their survival such as agriculture, construction, food cooking, mining and mineral processing. However, accidental NPs input to the environment has risen sharply and dramatically since the beginning of the Industrial Revolution in the mid-18th century and after due to manufacturing emissions, nuclear waste generation and the combustion of fossil fuels [33]. Diesel and automobile exhausts are the primary source of atmospheric nano-and microparticles in urban areas (2002). Since their productions are spontaneous, uncontrolled accidental nanoparticles are more likely to be polydisperse /heterogeneous and have irregular shapes. They contain sulfide, sulphate, nitrate, ammonium, organic carbon, elemental carbon and trace metals [33].

2.2.3 Engineered Nanoparticles

Nanostructured materials behave differently and are often superior in properties to bulk materials due to two primary factors: surface effects (enhanced surface phenomena caused by the very high surface area to volume ratio, most atoms are on or near the surface thus they can be weakly bonded and more reactive) and quantum effects (discontinuity in behavior caused by the confined delocalised electrons on the surface of the NPs due to their small dimensions) [34]. Those properties affect the overall physicochemical properties of the nanoscale materials in a way that they demonstrate novel and unprecedented properties.

Nanotechnology takes the advantage and exploits these aforementioned unique properties of nanoscale matter to design, translate and apply science to plenty of useful products [34]. The modern nanotechnology exploits the novel nanoscale properties of nanoparticles which has led to the production of a vast amount of engineered NPs. Engineered NPs are deliberately manufactured by human activities to serve for special purposes and they are different from both incidental NPs that are produced as a side product of human activity, for example, from industrial processes or transport and from natural NPs, for example, humic substances, produced from weathering, microbial action or chemical hydrolysis [35].
Engineered NPs can be divided into a number of classes and not as a single homogeneous group. Based on their core materials, these manufactured nanomaterials can be classified into organic and inorganic. Organic NPs can be further defined as fullerenes (C60 and C70 and derivatives) and carbon nanotubes (multi-walled or single walled CNTs), while inorganic NPs can be sub-divided into metal oxides (of iron, zinc, titanium, cerium etc), metals (mainly Silver, gold and copper) and quantum dots (such as cadmiumselenides) [36].

2.2.4 Production of Engineered Nanoparticles

Many different techniques have been developed and employed to generate metal nanoparticles, including silver nanoparticles. Those techniques for preparing nanoparticles have advanced rapidly over the recent decades and continue to evolve leading to more and improved control over the size and shape of the particles generated. Two fundamentally different approaches towards the controlled generation of nanostructures have evolved irrespective of the field or discipline [37]. The bottom up method (the chemical approach), where the atoms (produced from reduction of ions) are assembled to generate nanostructures, and the opposite approach, the top down method, also known as the physical method, where material is removed from the bulk material through grinding, milling, chemical methods or volatilisation of solid material followed by condensation of the vapour components, leaving only the desired nanostructures. Both approaches can be implemented in either gas, liquid, supercritical fluids, solid states, or in vacuum. Most of the manufacturers are interested in the ability to control one or more of the following aspects of the nanoparticles [38]: Particle size, Particle shape, Size distribution, Particle composition and Degree of particle agglomeration.
An important aspect of both approaches is the stabilisation of the particles to avoid aggregation and coalescence [38]. Stabilization can occur in many different ways but primarily aggregation is prevented by electrostatic repulsion or steric hindrance. Two of the most important issues confronting nanocrystal synthesis are obtaining a purposeful control over the nanocrystal mean size and routinely producing narrow size distributions [39].

2.3 Silver Nanoparticles

Among the different types of metal, nanoparticles silver is by far the most studied in the literature due to its wide and rapidly growing application in a number of scientific areas and in consumer products [43] and its known toxic effects on the environment and human health [44]. Its areas of application vary from textile engineering, catalysts [45], water treatment [46] and disinfecting to electronics and biotechnology [47].

2.3.1 Synthesis Silver of Nanoparticles

Colloidal silver is made by adding an excess of the reducing agent to silver salt. for example, by adding an excess of the reducing agent sodium borohydride, $NaBH_4$ to silver nitrate, $AgNO_3$, nanoparticles that are about 10-14 nm in diameter are produced [48].

$$AgNO_3 + NaBH_4 \rightarrow Ag + 1/2H_2 + 1/2B_2H_6 + NaNO_3 \qquad 2.2$$

2.3.2 Procedure for Synthesis of Silver Nanoparticles

A large excess of $NaBH_4$ was needed to reduce the ionic silver and to stabilize the silver nanoparticles that were formed. Different volumes of 0.001M silver nitrate were added drop wise (about 1 drop per second) to 30 ml of 0.002 M sodium borohydride solution that had been chilled in an ice bath. The reaction mixture was stirred vigorously on a magnetic stirrer. The solution turned to light yellow after the addition of 2 ml of silver nitrate and to brighter yellow when all of the silver nitrate had been added [48].

3.0 MATERIALS AND METHODS

3.1 The Chemical Synthesis of Silver Nanoparticles

Many chemical reduction methods have been used to synthesize silver nanoparticles from silver salts. Two main synthesis methods were adopted in the work of this thesis to synthesize silver nanoparticles, using sodium borohydride and trisodium citrate as reducing agents. These were primarily chosen on their possibilities to produce particles of different size regimes, and also because of their synthesis simplicity. In addition, variations in relative quantities of reagents, temperature and durations are performed. The concentration, size, shape and stability of the nanoparticles depend upon these conditions.

3.2 Chemicals

The chemicals used for the experiments include silver nitrate ($AgNO_3$, M = 169.87 g/mol); trisodium citrate ($Na_3C_6H_5O_7.2H_20$, M = 294.10 g/mol); Sodium borohydride ($NaBH_4$, M = 37.83).

3.3 Reduction of Silver Nitrate by Sodium Borohydride

One of most common methods to synthesize silver nanoparticles is by the use of sodium borohydride to reduce silver nitrate.

$$AgNO_3 + NaBH_4 \longrightarrow Ag + \frac{1}{2}H_2 + \frac{1}{2}B_2H_6 + NaNO_3 \quad (1)$$

3.4 Precaution

Great care has been taken during the synthesis to obtain colloidal silver, especially when dealing with a strong reducing agent such as sodium borohydride. Proper cleaning of laboratory equipment in advance is crucial, and the silver nitrate must by no means be exposed to the aqueous sodium borohydride except during the controlled addition.

3.5 Colloidal silver in water- Synthesis details

All equipment was cleaned thoroughly with distilled water. Reactions were done in a 250 ml conical flask. A given amount of aqueous silver nitrate contained in measuring cylinder was added drop wise to an aqueous silver nitrate, using a handheld dropper. The solution was stirred vigorously on a magnetic stir plate throughout the whole reaction (stirring speed = 600).

3.6.1 Experiment 1: Synthesis at Temperature of 27^0C

27.5 mg of $AgNO_3$ (molar weight = 169.87g/mol) was dissolved in 200 ml of deionised water to prepare a 0.8 mM aqueous silver nitrate solution. Analogously, a 1.6 mM aqueous sodium borohydride solution was made by dissolving 36.32 mg of $NaBH_4$ (molar weight = 37.83 g/mol) to 600 ml of deionised water. The compounds dissolved quite easily within 1 minute. The concentration of the sodium borohydride was twice that of the aqueous silver nitrate to ensure colloid stability fixed by the reaction stoichiometry.

The sodium borohydride is both a reducing agent and a stabilizing agent. Hence, different volumes of the precursors were reacted in order to get the ratio at which the sodium borohydride will effectively produce the nanoparticles and stabilize them. An overview of the compositions and durations of the experiment is given in **Table 1.**

Table 1: Details for the reaction of $AgNO_3$ with borohydride at 27°C.

Sample	$AgNO_3$: $NaBH_4$ ratio (ml)	Adding time (Minutes)
A1	14 : 50	2: 48
A2	10 : 50	2:02
A3	20 : 50	3:04
A4	5 : 50	1:13

All attempts to carryout synthesis above 20:50 $AgNO_3$:$NaBH_4$ ratio yielded instantly aggregated samples. However, only the sols listed in the above table were used in further characterization.

3.6.2 Experiment 2: Synthesis at Temperature of 4°C

The synthesis was performed with solutions of equal concentration as those of the first experiments. For these samples, the solutions were cooled down to 4°C. However, the initial volumes of the two solutions were half of the first experiment. This was done for better temperature stability at 4°C.

13.75 mg of $AgNO_3$ was dissolved in 100 ml of deionised water to prepare a 0.8 mM aqueous silver nitrate solution. Analogously, a 1.6 mM aqueous sodium borohydride solution was made by dissolving 18.16 mg to 300 ml of deionised water. Both solutions were kept in ice. The experiment was started when the temperature of the solutions stabilized at 4°C. An overview of the compositions and durations of the experiment is given in **Table 2.**

Table 2: Details for the reaction of silver nitrate with borohydride at 4°C

Sample	$AgNO_3$: $NaBH_4$ ratio (ml)	Adding time (Minutes)
B1	14 : 50	2:39
B2	10 : 50	2: 00
B3	20 : 50	3:14
B4	5 : 50	1:16

3.7 Reduction of $AgNO_3$ by Trisodium Citrate

Sodium borohydride is a very strong reducing agent. As a result of its reducing strength, a side reaction could jeopardize the synthesis of nanoparticles due to their sensitivity. The samples produced could also be harmful to the environment if not properly disposed. However, the synthesis of silver nanoparticles is achievable via an alternative route through an organic reducing agent sodium citrate which is less expensive and offers less potential hazards. However, vigorous conditions are needed for this synthesis.

3.7.1 Experiment 3: Using Trisodium Citrate

90 mg of $AgNO_3$ was dissolved in 500 ml deionized water to give 1.06 mM silver nitrate solution. Additionally, 1% (38 mM) trisodium citrate was easily prepared by dissolving 1g of $Na_3C_6O_7.2H_2O$ in 100 ml of deionized water. In a typical experiment, a beaker containing 50 ml of the aqueous silver nitrate was placed directly on top of a heater and heated up to boiling temperatures. Solution of trisodium citrate was added directly on top of the heater using a handheld dropper.

Different solutions ratios were used to investigate the effect on stability and the resulting nanoparticles characteristics. The details for the synthesis are given in **Table 3.**

Table 3: Details for the reaction of silver nitrate with trisodium citrate

Sample	$AgNO_3$:Citrate (ml)	Adding time (Minutes)	Post-addition stirring (Minutes)
C1	50: 25	4:19	5:20
C2	50: 35	5:15	5:20
C3	50: 45	7:13	6:05

3.8 Nanoparticles Characterisation

The characterization was performed by nano-drop with an inherent UV/Visible Spectrophotometer. The cover of the nano-drop was opened and the cuvette was filled with distilled water and blanked. About 1.5mil of a sample was poured into the cuvette, the cover was closed.

The optical properties (absorbance) of solution (sample) were evaluated with UV-VIS spectrophotometer at a wavelength range of: 200 nm –700 nm. It measures the intensity of light passing through a sample (I), and compares it to the intensity of light before it passes through the sample (I_o). The ratio I/I_o is called the transmittance, and is usually expressed as a percentage (%T). The absorbance, A, is based on the transmittance: $A = - \log (\%T)$. The scanning process was conducted using a computer system connected to the nano-drop. After the scanning is over a chromatograph of the absorbance versus the wavelength is displayed.

While scanning the sample a platform for naming the sample will pop out, this allows the user to name the sample. The chromatograph is stored inside the computer and subsequently, chromatographs of the same sample at variable times are added to the workbook.

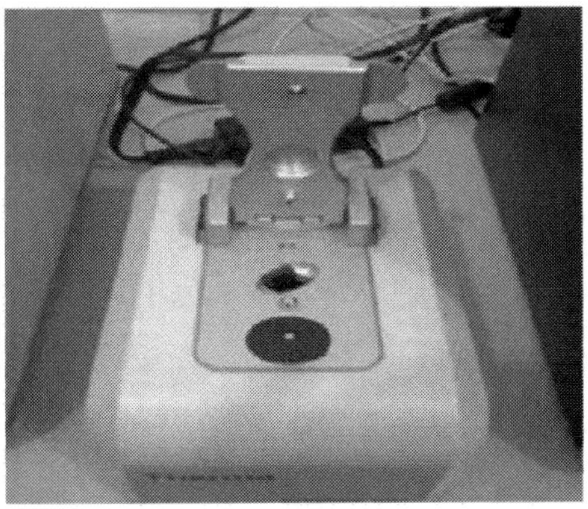

Figure 1: Nano-drop used for UV/Visible spectroscopy

Figure 2: Material Laboratory, North-East zonal Biotechnology Centre, University of Maiduguri, Nigeria

4.0 EXPERIMENTAL RESULTS

This chapter presents the result of all the experiments performed. Eleven samples were successfully synthesized at varying conditions and the spectrum of each sample was recorded three times: the day of synthesis, the third day and the seventh day. The quest was to find the best synthesis condition in terms of size, shape, concentration, size distribution, and stability.

4.1 Sample Synthesized from Sodium Borohydrate at 27^{0}C

Four samples (A1, A2, A3, A4) were synthesized. The synthesis details are given in Table 1.

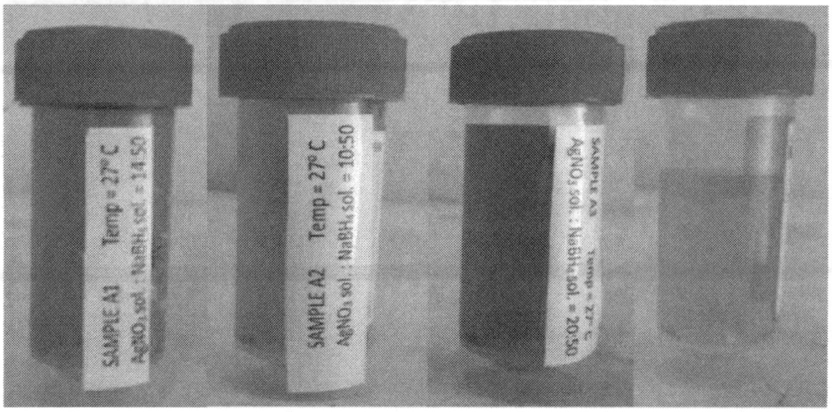

a b c d

Figure 3: Colloidal solutions of synthesis at 27^{0}C: (a) A1 (b) A2 (c) A3 (d) A4 [$NaBH_4$:$AgNO_3$ for A1, A2, A3, A4 are respectively 14:50, 10:50, 20:50 and 5:50]

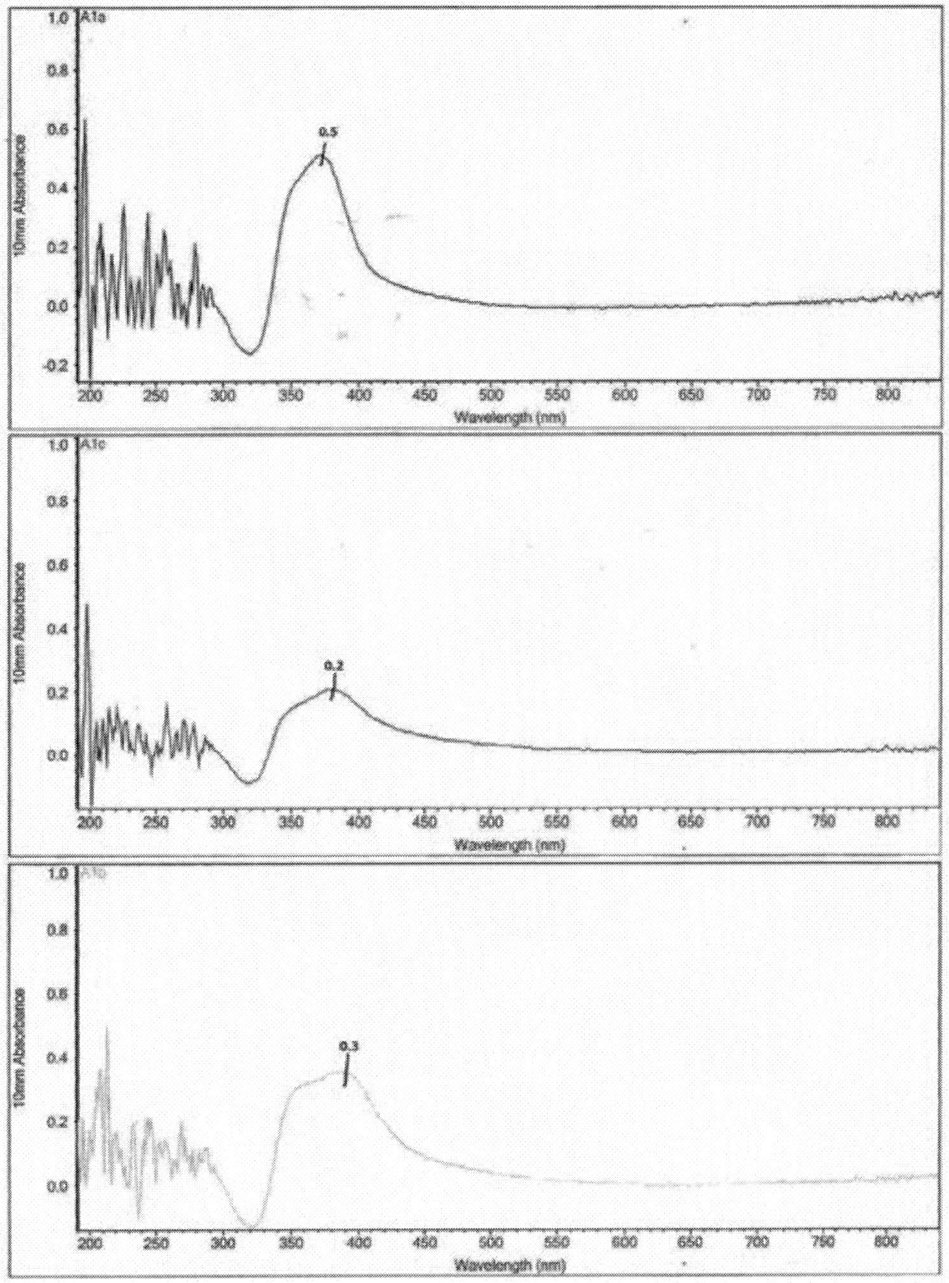

Fig. 4.1a: UV-Vis spectra of sample A1 [Time of synthesis: (a) 1 day (b) 3 days, (c) 7 days]

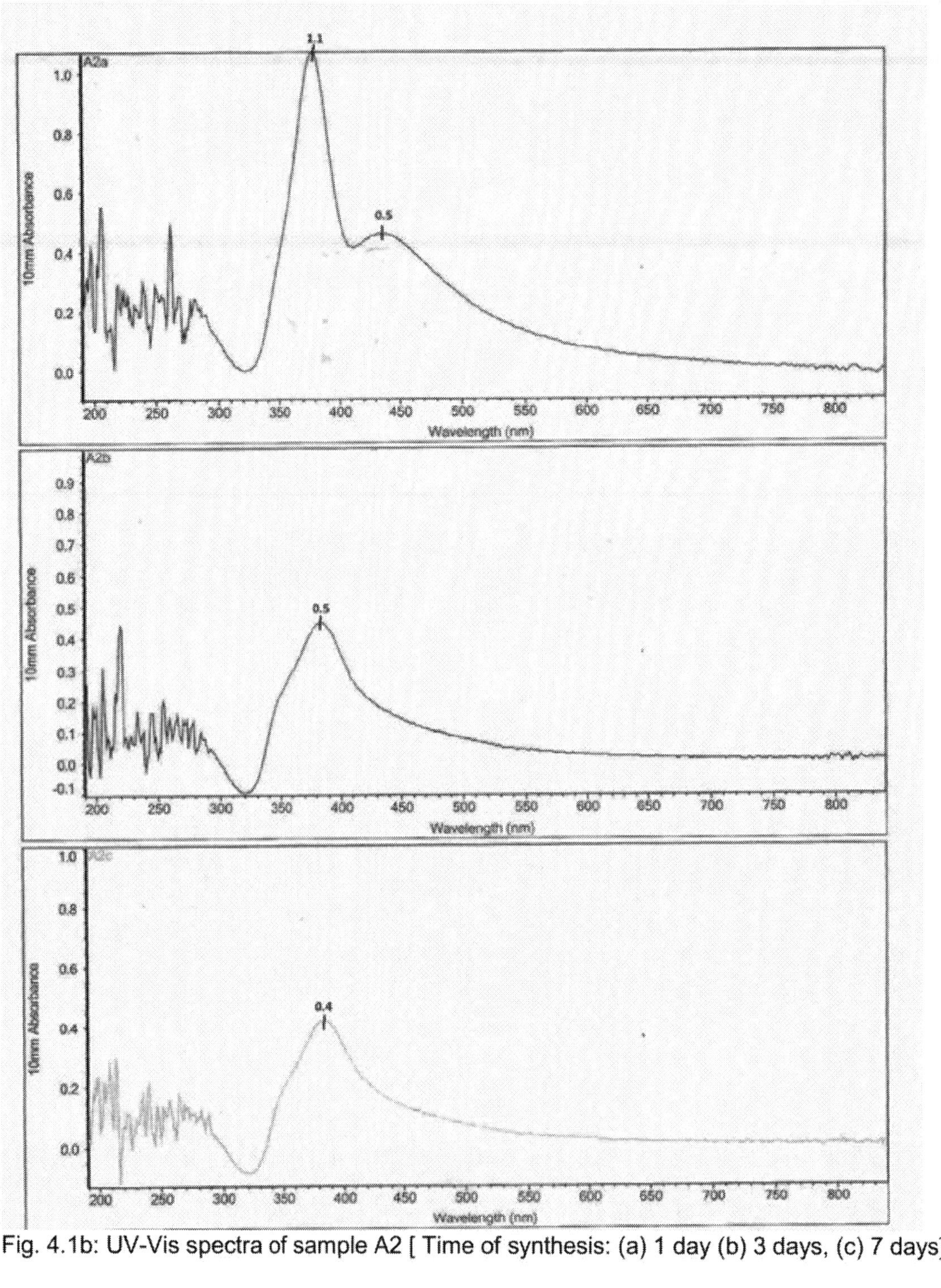

Fig. 4.1b: UV-Vis spectra of sample A2 [Time of synthesis: (a) 1 day (b) 3 days, (c) 7 days]

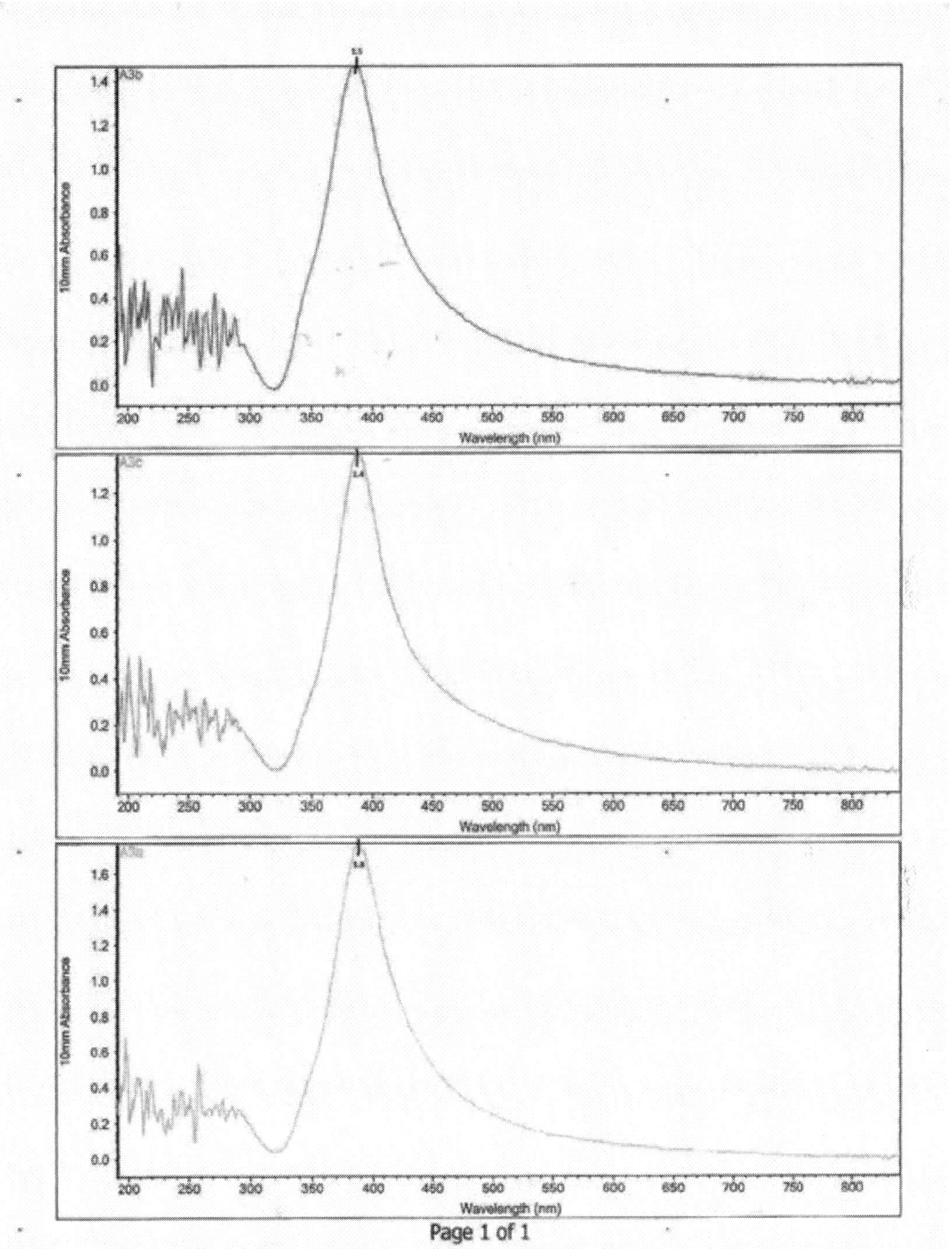

Fig. 4.1c: UV-Vis spectra of sample A3 [Time of synthesis: (a) 1 day (b) 3 days, (c) 7 days]

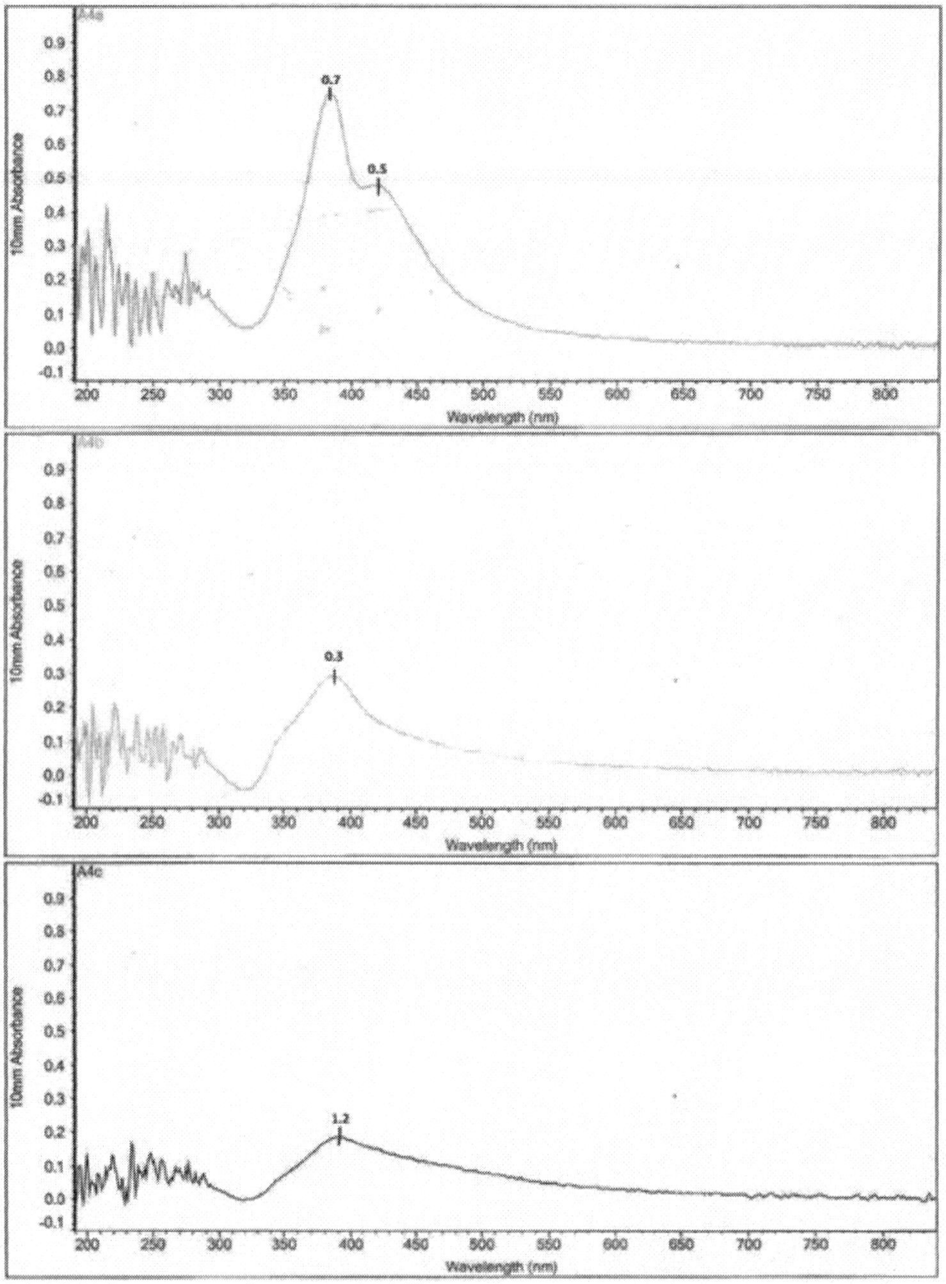

Fig. 4.1d: UV-Vis spectra of sample A4 [Time of synthesis: (a) 1 day (b) 3 days, (c) 7 days]

4.2 Sample Synthesized at 4^{0}C

Four samples (B1, B2, B3, and B4) were synthesized at 4^{0}C and their spectra were obtained at three different times (the first day, the third day and the seventh day). The details of the synthesis are given in Table 3.2. Larger ratios was attempted, but all attempts yielded an aggregated solution. Only the successfully synthesized solutions were characterized.
Figure 4.2 shows the appearance of the samples.

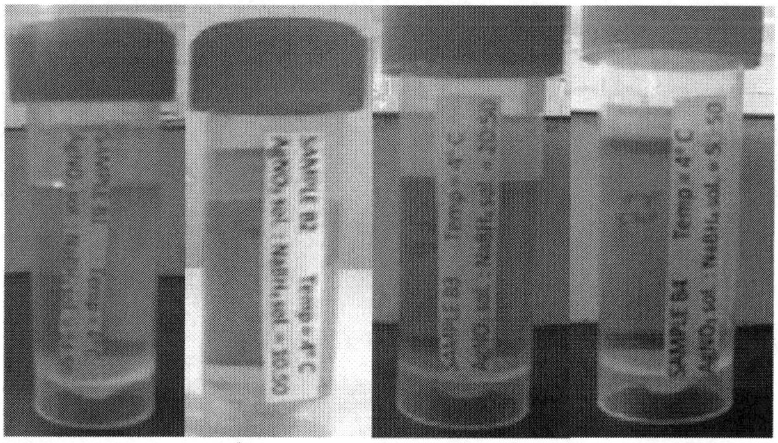

a b c d

Fig. 4.2: Colloidal solutions of synthesis at $4^{0}C$ (a) B1 (b) B2 (c) B3 (d) B4 [$NaBH_4$:$AgNO_3$ for B1, B2, B3, B4 are respetively 14:50, 10:50, 20:50 and 5:50]

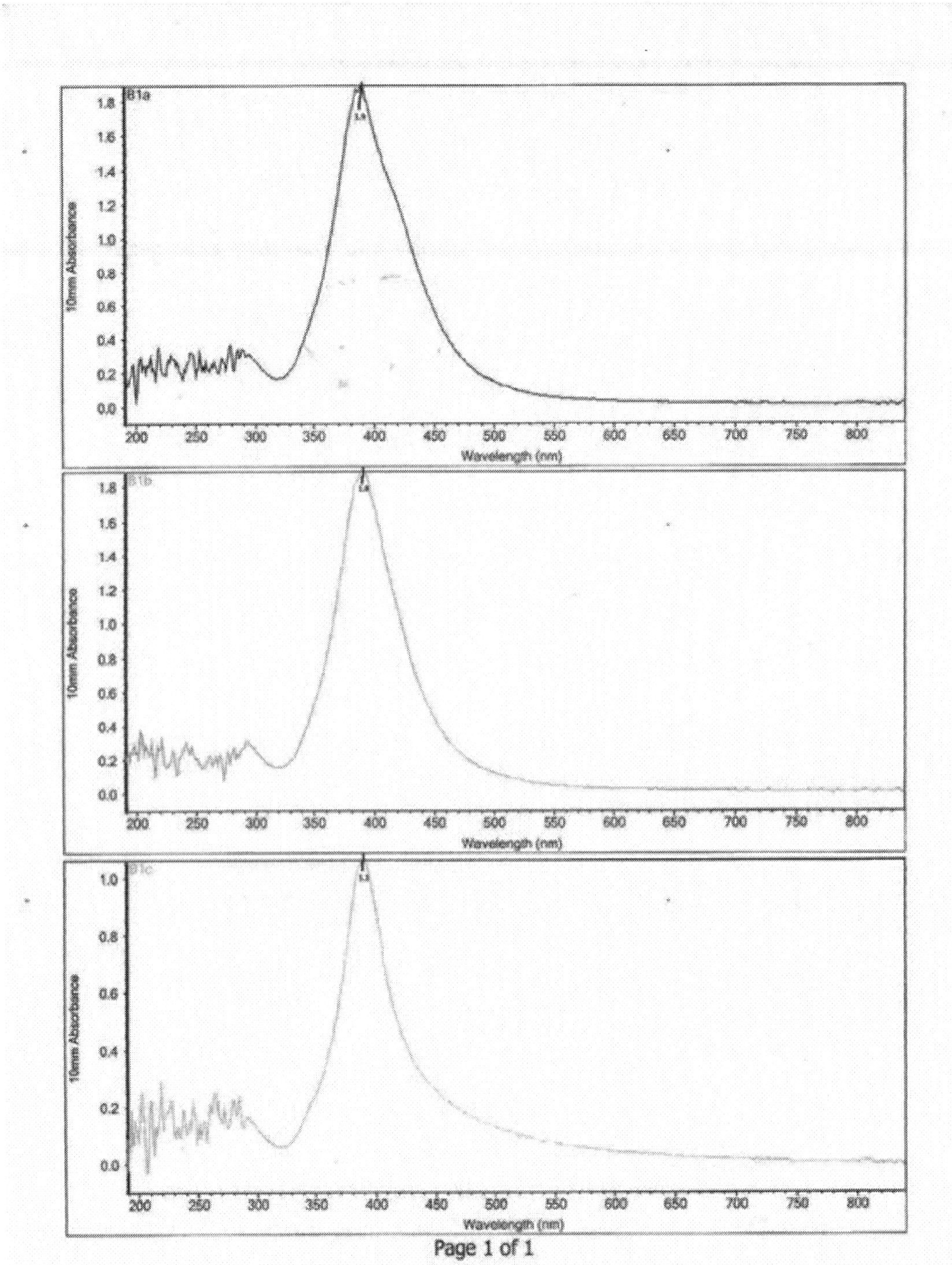

Fig. 4.2a: UV-Vis spectra of sample B1 [Time of synthesis: (a) 1 day (b) 3 days, (c) 7 days]

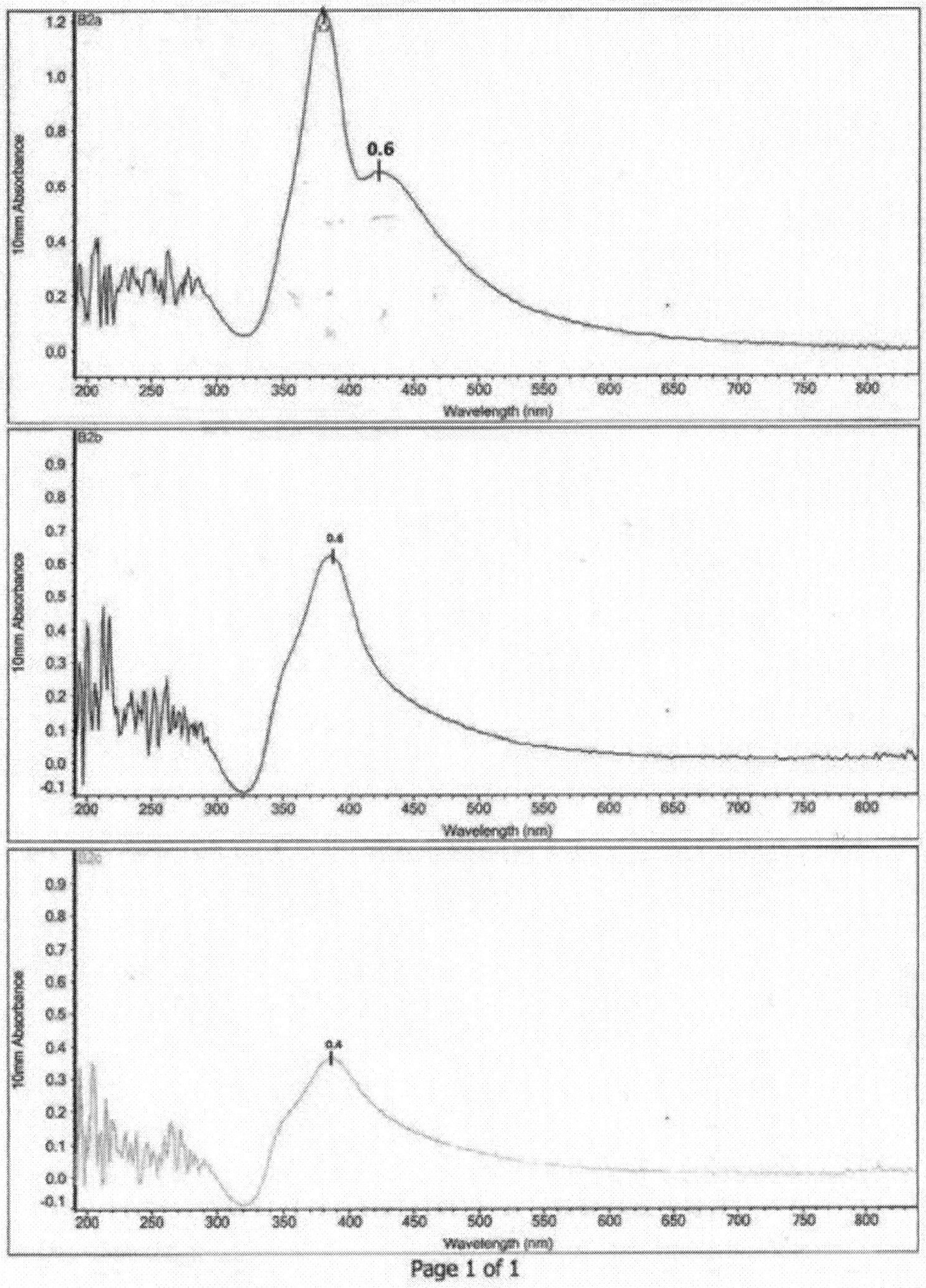

Fig. 4.2b: UV-Vis spectra of sample B2 [Time of synthesis: (a) 1 day (b) 3 days, (c) 7 days]

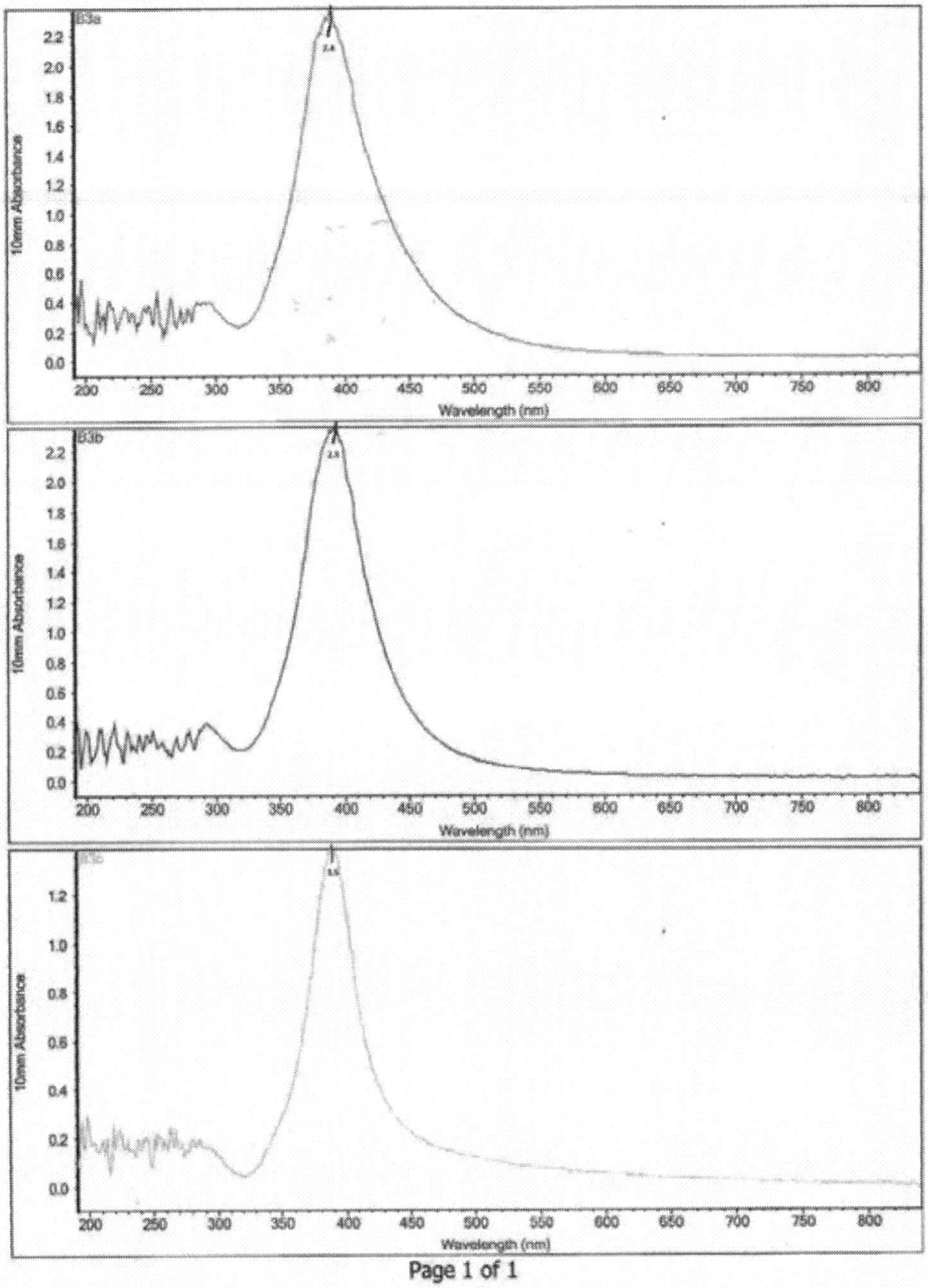

Fig. 4.2c: UV-Vis spectra of sample B3 [Time of synthesis: (a) 1 day (b) 3 days, (c) 7 days]

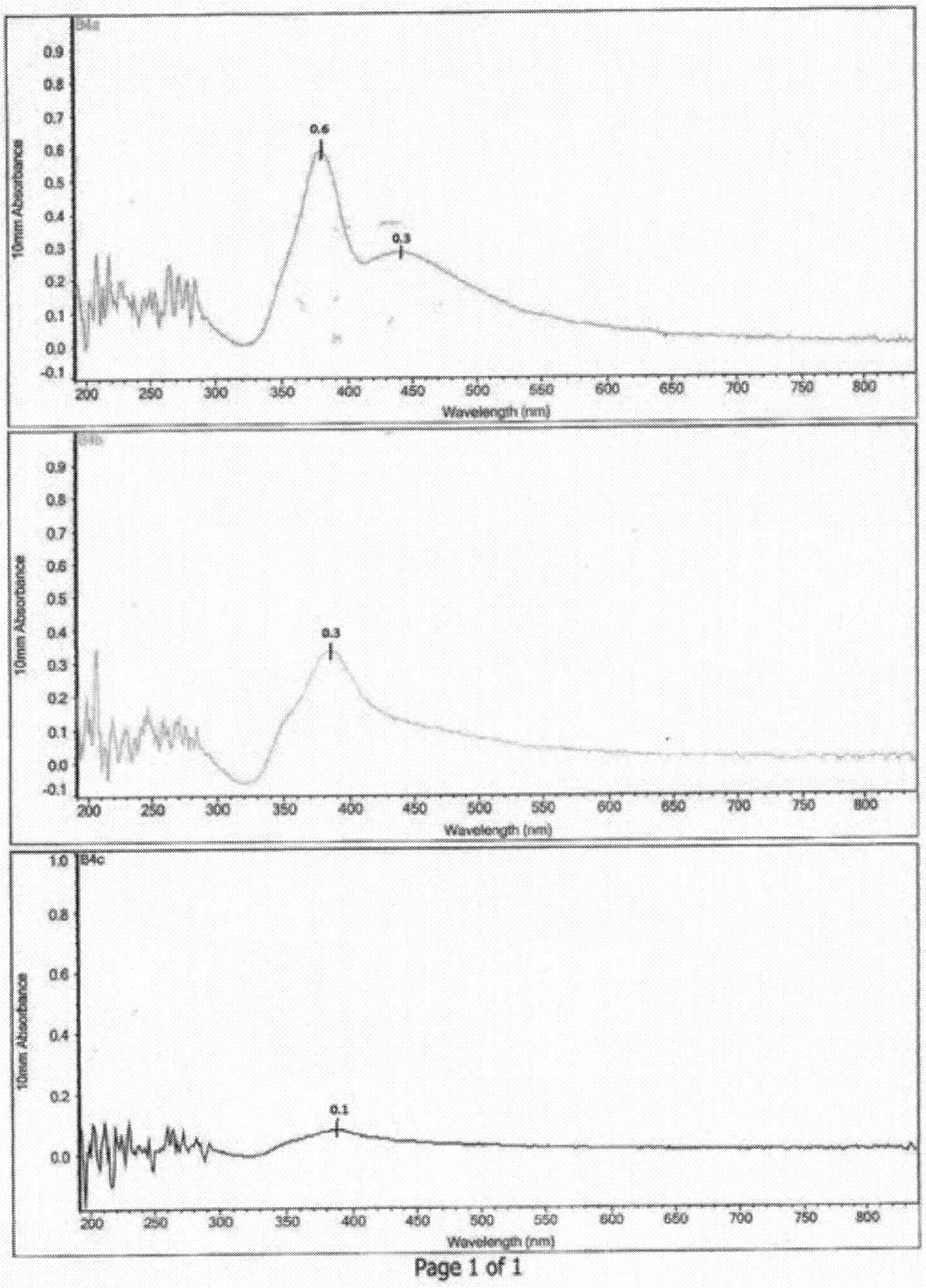

Fig. 4.2d: UV-Vis spectra of sample B4 [Time of synthesis: (a) 1 day (b) 3 days, (c) 7 days]

4.3 SAMPLES SYNTHESIZD BY SODIUM CITRATE

Three samples (C1, C2 and C3) were synthesized by reduction of silver salt ($AgNO_3$) using trisodium citrate. The reaction was carried out at the tempreture of boiling water. The solution was vigorously stirred during and after addition of the silver salt to citrate salt. The details of the experiment is given in Table 3.3

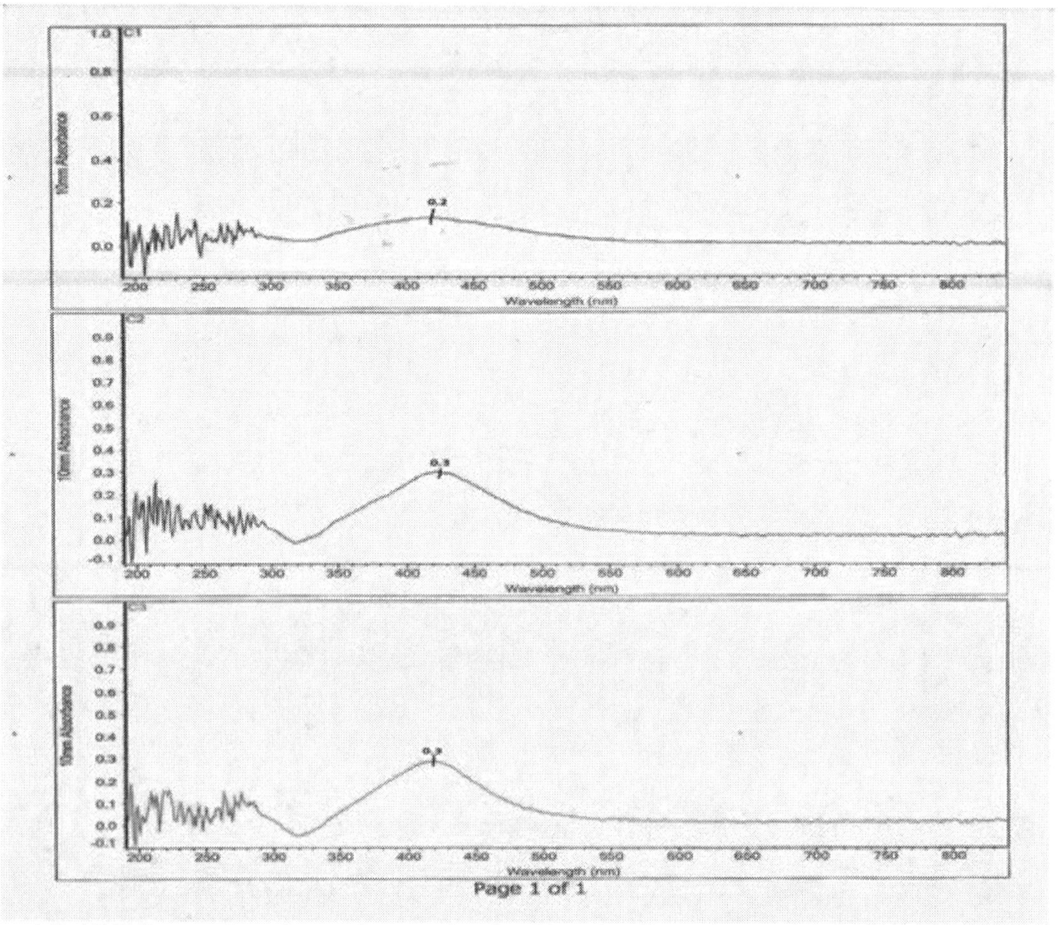

Fig. 4.3: UV-Vis spectra of samples synthesized by trisodium citrate: (a) C1 (b) C2, (c) C3

5.0 DISCUSSION OF RESULTS

In order to find out the effect of various synthesis conditions on the resulting nanoparticles, the nanoparticles were characterized. For the characterization of synthesized NPs, special attention was devoted on the quality of the product of the synthesis methods by comparing the properties of the nanoparticles in terms of size, size distribution, monodispersity, sphericity, concentration and stability of the NPs.

5.1 Nanoparticles and UV-Visible Spectroscopy

The dispersions of silver nanoparticles display intense colors due to the Plasmon resonance absorption. Metals with free electrons possess Plasmon resonances in the visible spectrum, which give rise to such intense colours [61]. These properties are mainly observed in Au, Ag, and Cu because of the presence of free conduction electrons. The electric field of the incoming radiation induces the formation of a dipole in the nanoparticle. A restoring force in the nanoparticle tries to compensate for this, resulting in a unique resonance wavelength. Therefore metallic nanoparticles have characteristic optical absorption spectrum in the UV-Visible region [62]. This makes UV-Visible spectroscopy an indispensable tool for characterization of nanoparticles.

5.2 Synthesis and Growth of $AgNO_3$ Nanoparticles

The synthesis of silver nanoparticles was carried out using three different approaches. The first samples with codes A1, A2, A3, A4 (see Table 1 for synthesis details) were synthesized with sodium borohydride at room temperature. When the reducing agent ($NaBH_4$) was introduced to the solution of $AgNO_3$, the initial colourless solution turned instantaneously to yellow, which indicates that the silver ions were spontaneously reduced to silver atoms, which in turn, simultaneously clumped into silver nanoparticles, as reported [63]. The yellow colour formed within 2 minutes remained unchanged, this indicates the formation of stable silver nanoparticles [63]. In similar manner, the colour change and the growth of samples B1, B2, B3, B4 (see Table 2 for synthesis details) synthesized with sodium borohydride at room temperature manifested the same pattern as the above, although their formation was less drastic which indicates that the reaction rate was faster at higher temperatures. However, the colour changes of samples C1, C2, C3 using sodium citrate were observed after heating the solution to about 100^{0}C with vigorous stirring and waiting to about 15 minutes. Total reaction time of these last samples was about 30 minutes. This slow change can be attributed to the weak reducing effect of the citrate salt [64].

5.3 PRESENCE OF SILVER NANOPARTICLES

In all the experiments summarized above and described in Chapter three and subsections, the colours of the solutions have changed from colourless to yellow, due to silver ions from precursor to silver nanoparticles, which is characteristic formation of silver in nanoscale [62, 63]. The colour change manifested is caused by surface plasmon resonance (SPR) of the forming silver NPs [62, 63]. Therefore, the presence of silver NPs in the final suspensions was further confirmed by Uv-Vis absorbance measurement which shows the NPs localised surface Plasmon resonance (LSPR) behaviours. The spectra are given in chapter five and clearly indicate the characteristic sharp absorbance peak around 370-440 nm for all samples, which is a characteristic of silver nanoparticles [65].
Electromagnetic radiation looks white when all wavelengths of radiation from 350 nm to 750 nm are present. White light is composed of various hues of red, orange, yellow, green, blue, indigo, and violet light. When a specie absorbs selected portions of the visible light spectrum, color results. For example, the solution of silver nanoparticles mentioned above is a yellow solution that strongly absorbs light at around 400 nm.

Table 4: Absorbance and Complementary Colour

Wavelength	Absorbed Colour	Complementary Colour
650-780	red	blue-green
595-650	orange	greenish bue
560-595	yellow-green	purple
500-560	green	red-purple
490-500	bluish green	red
480-490	greenish blue	orange
435-480	blue	yellow
380-435	violet	yellow-green

Source: Owen, (1999)

From Table 4, the yellow color of the solution is the result obtained when the violet component of white light is removed. The color of the solution, therefore, represents a composite of all colors transmitted (i.e., not absorbed) by the particles. It is often possible to determine the color (or wavelength) of light absorbed by a specie based on the color of the solution and vice versa. In simple cases, where a solution absorbs light in only one region of the visible spectrum, the color of the solution is complementary to the color of light absorbed by the solution.

5.4 Particle Size Characterisation

Table 5: Spectroscopy and Size of Nanoparticle

Particle size nm	Λ_{max} nm	PWHM nm
10-14	385-400	50-70
35-50	420	100-110
60-80	438	140-150

(International Journal of Engineering Research & Technology, 2013)

The position of the maximum absorbance measured with Uv-Vis depends on a number of factors including the shape, size and the surrounding medium. As a general rule, as the particles become larger, the absorption maximum shifts to longer wavelength [66].
All samples synthesized from sodium borohydride irrespective of synthesis temperature [Fig 4.1 (a, b, c and d) and Fig 4.2 (a, b, c and d)] showed absorption maximum at wavelength of around 387 nm, as reported by other researchers [67]. However, samples synthesized from sodium citrate (Fig 4.3) showed absorption maximum at around 420 nm. Accordingly, from Table 5.2, the sizes all samples synthesized from sodium borohydride were around 12 nm as reported by other researches [67]. Also, nanoparticles synthesized from trisodium citrate were around 35 nm, as also reported [68].
Similarly, symmetrical particles absorb at longer wavelength [66] which indicates that particles from sodium borohydride were perfectly spherical and particles from sodium borohydride were roughly spherical.
However, sodium borohydride samples synthesized with lower $NaBH_4$:$AgNO_3$ solution ratios [A2 and B2 with $NaBH_4$:$AgNO_3$ (10:50); A4 and B4 with $NaBH_4$:$AgNO_3$ (5:50)] showed double peaks at 387 nm and 440 which indicates distribution of two nanoparticles of sizes 12 nm and 37 nm.

5.5 Stability

The optical properties of silver nanoparticles change when particles aggregate and the conduction electrons near each particle surface become delocalized and are shared amongst neighbouring particles. When this occurs, the surface Plasmon resonance shifts to lower energies, causing the absorption and scattering peaks to red-shift to longer wavelengths [66].
To monitor the stability of the silver colloids, we have measured the absorption of the colloids at different periods of time (Day 1, Day 3, and Day 7).
The UV-VIS spectrums recorded after different time intervals shown Figure 4.1 (a, b, c and d) and Figure 4.2 (a, b, c and d) are summarized in Table 5.3a and Table 5.3b respectively.

Table 5.3a: Summary of absorbance and corresponding wavelength recorded at three different times for samples prepared from sodium borohydride at 27^{0}C

	Time of Synthesis					
	Day1		Day2		Day3	
Sample	**Maximum Absorbance**	**λ_{max}**	**Maximum Absorbance**	**λ_{max}**	**Maximum Absorbance**	**λ_{max}**
A1	**0.5**	**385**	**0.4**	**387**	**0.2**	**388**
A2	**1.1**	**385**	**0.5**	**388**	**0.5**	**388**
A3	**1.8**	**386**	**1.5**	**387**	**1.4**	**387**
A4	**0.7**	**386**	**0.3**	**388**	**0.2**	**389**

Table 5.3b: Summary of absorbance and corresponding wavelength recorded at three different times for samples prepared from sodium borohydride at 4^0C

	Time of Synthesis					
	Day1		Day2		Day3	
Sample	Maximum Absorbance	λ_{max}	Maximum Absorbance	λ_{max}	Maximum Absorbance	λ_{max}
B1	1.9	386	1.8	387	1.1	387
B2	1.2	387	0.6	388	0.5	388
B3	2.4	387	2.2	387	1.5	387
B4	0.6	386	0.3	388	0.1	390

The extinction peaks for the samples have shifted to longer wavelength, decreased in intensity and have broadened. However, this change was more drastic for samples synthesized at 27^0C. This was also observed from the colour changes of the samples as the yellow darkens, turns violet, then grayish as the particles settle out within days of synthesis. However, samples synthesized with higher $NaBH_4$:$AgNO_3$ (B3, A3) showed an appreciable stability as their spectrum stabilized and their colour persisted for weeks after synthesis.

5.6 Effects of Difference Synthesis Condition

For the synthesis of silver nanoparticles, multi-approach was applied so as the see the effects of different synthesis conditions on the resulting nanoparticles. These conditions are examined in the coming sub sections.

5.6.1 Effect of Variable $NaBH_4$:$AgNO_3$ Solution Ratios

Before reducing agent is added to the silver solution, there are only silver ions. The reducing agent, sodium borohydride, will force silver atoms to form in solution. The concentration of silver atoms will rise rapidly until supersaturation is achieved, at which aggregation will occur [69]. The concentration of particles will depend on the number of nucleation sites: the more the particles [69]. The amount of nucleation sites is dependent on the amount of reducing agent. As the reducing agent is increased, there will be more nuclei and thus more nanoparticles. With too much reducing agent, the nucleation sites will form at different times and destabilize much quicker [69].

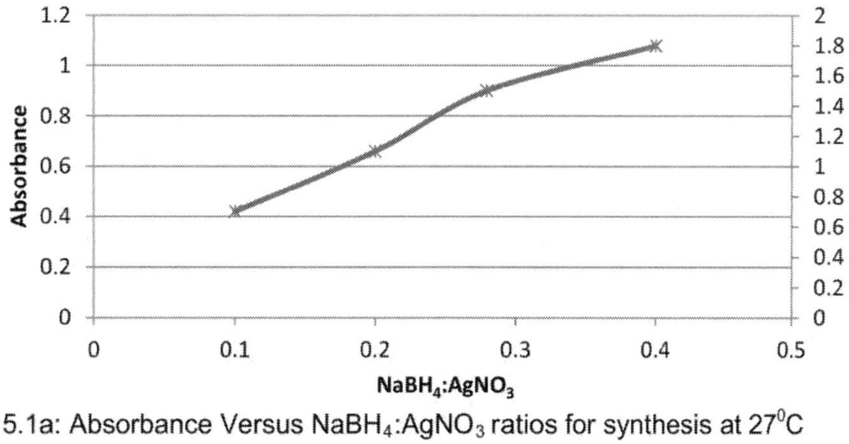

Fig 5.1a: Absorbance Versus $NaBH_4$:$AgNO_3$ ratios for synthesis at 27^0C

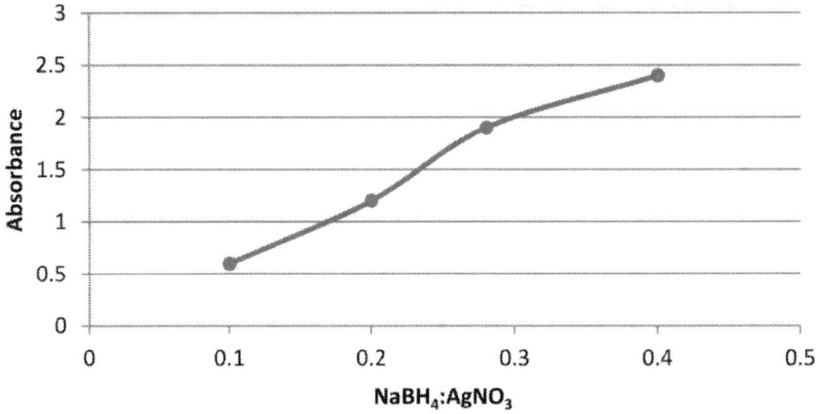

Fig 5.1b: absorbance versus $NaBH_4$:$AgNO_3$ ratios for synthesis at 4^0C

The absorbance of samples increase steadily with increase in $NaBH_4$:$AgNO_3$ for both synthesis at 4^0C and 27^0C as shown in fig 5.4a and fig 5.4b. This shows that conversion increase with an increase in amount of the reducing agent. However, an excess of the nuclei will be generated when $NaBH_4$ added is too high, which results in agglomeration. This was confirmed in the experiment as all attempts to perform the synthesis at $NaBH_4$:$AgNO_3$ > 20:50 yielded an instantly aggregated solution.

5.6.2 Effect of Synthesis Temperature

Chemical reactions are faster at higher temperature. However, due to a very high reducing effect of $NaBH_4$, impurities can easily be easily brought into the system as product of side reactions.

Samples synthesized at 4^0C were clear yellow (Fig 4.1) while samples synthesized at 27° were yellowish-brown (Fig 4.2) which indicates the possibility of presence of impurities.

Fig 5.5 shows that the concentration of samples synthesized at 4^0C is higher than that of samples synthesized at 27^0C. However, the effect of synthesis temperature is more drastic at higher $NaBH_4$:$AgNO_3$.

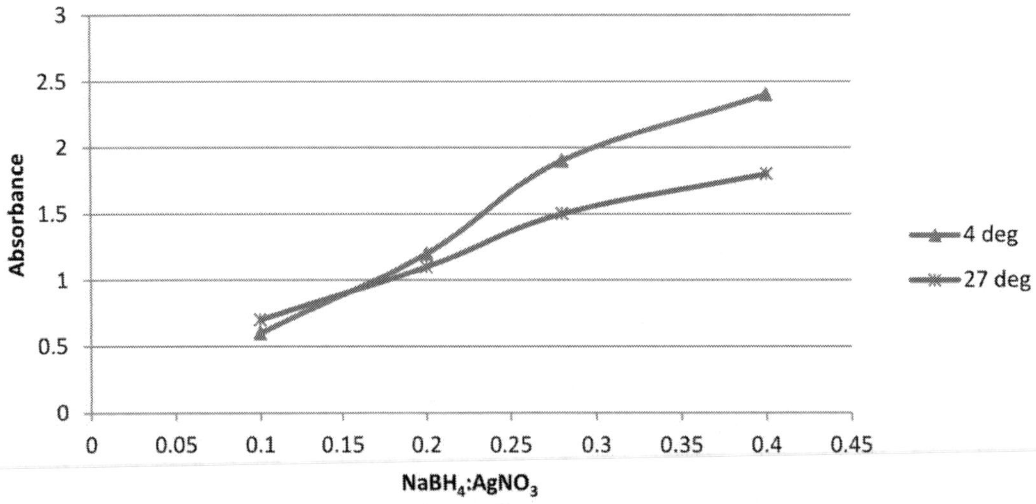

Figure 5.2: Effect of synthesis temperature

To study the effect of temperature on concentration and stability of the nanoparticles, Fig 5.6a and Fig 5.6b give an overview of synthesis at 4^0C and 27^0C.

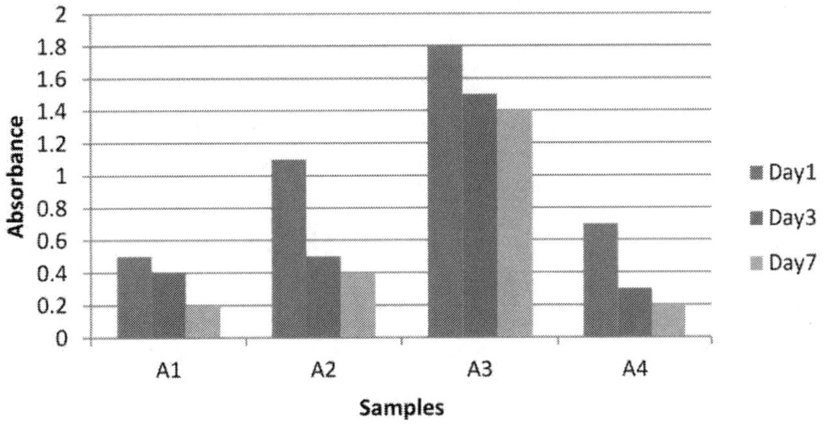

Figure 5.3a: An Overview of Synthesis at 27^oC

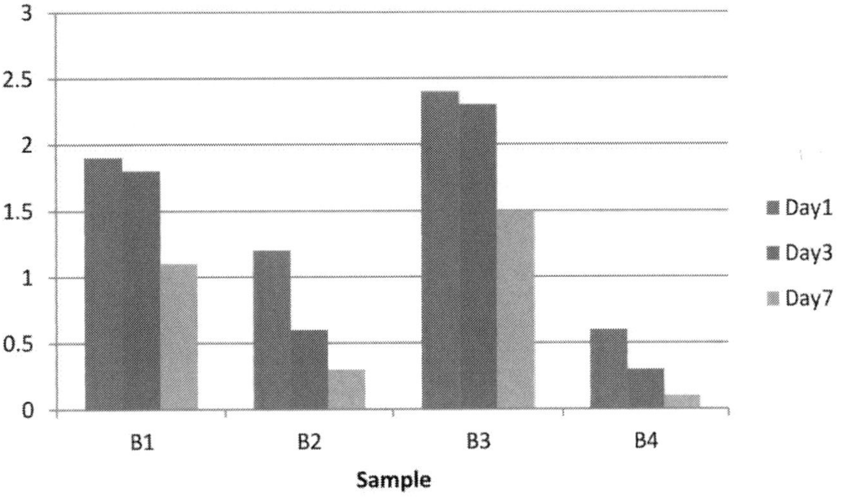

Figure 5.3b: An Overview of Synthesis at 4^oC

It can be seen that samples prepared at 4^0C were more concentrated and stable than samples prepared at 27^0C.

5.5.3 Effects of Using Trisodium Citrate As Reducing Agent

Three samples were successfully prepared from sodium citrate (the details of the synthesis are given in Table 3.3). The reaction lasted for about 15 minutes with vigorous stirring and heating. The resulting solutions were cloudy yellow which showed that the conversion of silver ions to silver nanoparticles were moderate. This was also confirmed by their spectra (Fig 4.3): the peaks were less intense, broader than that of samples from sodium borohydride.

Similarly, the sizes of particles were bigger (about 35 nm) and structurally less regular than particles obtained from sodium from borohydride. However, the samples was found to be stable under atmospheric conditions, this could be attributed to the repelling actions of citrate ions [66].

6.0 CONCLUSION

Silver nanoparticles were synthesized by reducing Ag^+. Two reducing agents were used. In the first and second experiment, sodium borohydride was used at synthesis temperature of 27^{0}C and 4^{0}C respectively. In the last experiment, sodium borohydride was replaced with trisodium citrate. In all the three experiments, variable reactants' volumes were used. The absorption spectrum of each sample synthesized was determined using UV-Visible spectrophotometer at three different times. Below were findings of the research:

During the synthesis, the first sign for the production of the nanoparticles was manifested in colour change from colourless solution to yellow. The formation of silver nanoparticles was confirmed by the characteristic surface Plasmon resonance for silver which is between 370-450 nm. Synthesis of silver nanoparticles via sodium borohydride take 1-2 minutes due to high reducing strength of sodium borohydride. Irrespective of synthesis temperature, colloids produced via sodium borohydride have absorption peak at 387 nm, which suggests that the particles were 12 nm in diameter. Low reaction temperature is helpful in optimum nanoparticles formation. Low-size, uniform silver nanoparticles with low polydispersity are obtainable at higher $NaBH_4$:$AgNO_3$ volume ratios. Nanoparticles synthesized from sodium citrate showed maximum absorption at 440 nm, which indicated the presence of nanoparticles of sizes around 35 nm. Nanoparticles synthesized by sodium borohydride were smaller, more concentrated and less distributed than nanoparticles synthesized by trisodium citrate. The best synthesis conditions for synthesis are: Using sodium borohydride as reducing agent at low temperature and high $NaBH_4$:$AgNO_3$ volume ratios.

REFERENCE

1. Kiss, L. B.; Söderlund, J.; Niklasson, G. A.; Granqvist, C. G. (1999). "New approach to the origin of lognormal size distributions of nanoparticles." *Nanotechnology* **10**: 25.

2. Alemán, J.; Chadwick, A. V.; He, J.; Hess, M.; Horie, K.; Jones, R. G.; Kratochvíl, P.; Meisel, I.; Mita, I.; Moad, G.; Penczek, S.; Stepto, R. F. T. (2007). "Definitions of terms relating to the structure and processing of sols, gels, networks, and inorganic-organic hybrid materials (IUPAC Recommendations 2007)."*Pure and Applied Chemistry* **79** (10).

3. Vert, Michel; Doi, Yoshiharu; Hellwich, Karl-Heinz; Hess, Michael; Hodge, Philip; Kubisa, Przemyslaw; Rinaudo, Marguerite; Schué, François (2012). "Terminology for biorelated polymers and applications (IUPAC Recommendations 2012)." *Pure and Applied Chemistry* **84** (2): 377–410.

4. Whetten, R. L.; Khoury, J. T.; Alvarez, M. M.; Murthy, S.; Vezmar, I.; Wang, Z. L.; Stephens, P. W.; Cleveland, C. L.; Luedtke, W. D.; Landman, U. Adv. Mater.1996, 8, 428−433.

5. Zheng, J.; Nicovich, P. R.; Dickson, R. M. Annu. Rev. Phys. Chem. 2007, 58, 409−431.

6. Christof, M. N. (2001). "Nanoparticles, Proteins, and Nucleic Acids: Biotechnology Meets Materials Science." *Angewandte Chemie International Edition*, 40, 4128-4158.

7. Shanmin, G. & et al. (2003). "Preparation of $CuAlO_2$ nanocrystalline transparent thin films with high conductivity." *Nanotechnology*, 14, 538.

8. Shafirovich, E.; Diakov, V. A. (2006). "Combustion of novel chemical mixtures for hydrogen generation." *Combustion and Flame*, 144, 415-418.

9. Paolo Pengo, S.P.; Marino, Battagliarin; Lucia, Pasquato; Paolo Scrimin (2003). "Synthesis, characterization and properties of water-soluble gold nanoparticles with tunable core size." *Journal of Materials Chemistry*, 2471 -2478.

10. Chattopadhyay, D.P.; Patel B.H (2009). "Improvement in physical and dyeing properties of natural fibres through pre-treatment with silver nanoparticles." *Indian Journal of Fibre & Textile Research*, 34, 368-373.

11. Chattopadhyay, D.P.; Patel B.H. (2010). "Effect of nanosized colloidal copper on cotton fabric." *Journal of Engineered Fiber Fabrics*, 1-6.

12. Jana, N.R.; Wang Z.L., Sau T.K. (2000). "Seed-mediated growth method to prepare cubic copper nano particles." *Current Science*, 79, 1367-1370.

13. Owen, Tony (1996). "Fundamentals of Modern UV-Visible Spectroscopy." *Hewlett-Packard*.

15 Daniel, M.C.; Astruc , D. (2004) *Chem. Rev.* , 104 , 293 – 346 .
16- Helcher, H.H. (1718). "Aurum Potabile Oder Gold Tinstur." J. *Herbord Klossen,* Germany
17. Helcher, H.H. (1718). "Aurum Potabile Oder Gold Tinstur." *J. Herbord Klossen*, Germany.
18. Faraday, M. (1857) *Philos. Trans. R. Soc.* London , 147 , 145 – 181 .
19. Satoshi, Horikoshi; and Nick, Serpone (2013). "Microwaves in Nanoparticle Synthesis." First Edition, *Wiley-VCH Verlag GmbH & Co*, p3.
20. Jin, R.; Cao, Y.; Mirkin, C.A.; Kelly, K.L.; Schatz, G.C. ; Zheng, J.G. (2001) *Science* , 294 , 1901 – 1903 .
21. For the details about The New York Times article of February 22, 2005 see: http://www.nytimes.com/imagepages/2005/02/21/science/20050222_NANO1_GRAPHIC.html.
22. Onoda, G.Y.; Hench, L.L. (1979). "Ceramic Processing Before Firing." *New York: Wiley & Sons.* ISBN 0-471-65410-8.
23. Aksay, I.A., Lange, F.F., Davis, B.I. (1983). "Uniformity of Al_2O_3-ZrO_2 Composites by Colloidal Filtration." *J. Am. Ceram. Soc.* **66** (10): C–190.
24. Fahlman, B. D. (2007). *Materials Chemistry*. Springer. pp. 282–283. ISBN 1-4020-6119-6.
25. Taylor, Robert A; Otanicar, Todd; Rosengarten, Gary (2012). "Nanofluid-based optical filter optimization for PV/T systems." *Light: Science & Applications* **1** (10): e34.
26. Taylor, Robert; Coulombe, Sylvain; Otanicar, Todd; Phelan, Patrick; Gunawan, Andrey; Lv, Wei; Rosengarten, Gary; Prasher, Ravi; Tyagi, Himanshu (2013). "Small particles, big impacts: A review of the diverse applications of nanofluids." *Journal of Applied Physics* **113**: 011301. doi:10.1063/1.4754271.
27. Taylor, Robert A; Otanicar, Todd; Rosengarten, Gary (2012). "Nanofluid-based optical filter optimization for PV/T systems." *Light: Science & Applications* **1** (10): e34.
28. Taylor, Robert A; Phelan, Patrick E; Otanicar, Todd P; Adrian, Ronald; Prasher, Ravi (2011). "Nanofluid optical property characterization: Towards efficient direct absorption solar collectors." *Nanoscale Research Letters* **6** (1): 225
29. Alemán, J.; Chadwick, A. V.; He, J.; Hess, M.; Horie, K.; Jones, R. G.; Kratochvíl, P.; Meisel, I.; Mita, I.; Moad, G.; Penczek, S.; Stepto, R. F. T. (2007). "Definitions of terms relating to the structure and processing of sols, gels, networks, and inorganic-organic hybrid materials (IUPAC Recommendations 2007)." *Pure and Applied Chemistry* **79** (10): 1801
30- Nowack, B.; Bucheli, T. D. (2007). "Occurrence, behavior and effects of nanoparticles in the environment." *Environmental Pollution*, 150, 5-22.
31- Theng, B.K.G.; Yuan, G. (2008). "Nanoparticles in the Soil Environment." *Elements*, 4, 395-399.
32. Sapkota, A.; Williams, D. A.; Buckley, T. J. (2005). "Tollbooth Workers and Mobile Source-Related Hazardous Air Pollutants: How Protective Is the Indoor Environment?" *Environmental Science & Technology,* 39, 2936-2943.
33. Wiesner, M. R.; Lowry, G. V.; Jones, K. L., Hochella, J. (2009). "Decreasing Uncertainties in Assessing Environmental Exposure, Risk, and Ecological Implications of Nanomaterial." *Environmental Science & Technology*, 43, 6458-6462.
34. Roduner, E. (2006). "Size matters: why nanomaterials are different." *Chemical Society Reviews*, 35, 583-592.
35. Lead, J. R.; Wilkinson, K. J. (2006). "Aquatic Colloids and Nanoparticles: Current Knowledge and Future Trends." *Environmental Chemistry*, 3, 159-171
36. Ju-nam, Y.; Lead, J.R. (2008). "Manufactured nanoparticles: An overview of their chemistry, interactions and potential environmental implications." *Science of the Total Environment*, 400, 396-414.
37. Shenhar, R.; Rotello, V.M. (2003). "Nanoparticles: Scaffolds and Building Blocks." *Accounts of Chemical Research*, 36, 549-561.
38. Ju-nam, Y.; Lead, J. R. (2008). "Manufactured nanoparticles: An overview of their chemistry, interactions and potential environmental implications." *Science of The Total Environment*, 400, 396-414.
39. Shields, S. P.; Richards, V. N.; Buhro, W. E. (2010). "Nucleation Control of Size and Dispersity in Aggregative Nanoparticle Growth. A Study of the Coarsening Kinetics of Thiolate-Capped Gold Nanocrystals. " *Chemistry of Materials*.
40. Faraday, M. (1857). "The Bakerian Lecture: Experimental Relations of Gold (and Other Metals) to Light." *Philosophical Transactions of the Royal Society of London*, 147, 145-181
41. Turkevich, P.C.; J., Hiller (1951). "A study of the Nucleation and Growth Processes in the Synthesis of Colloidal Gold." *Discussions of the Faraday Society*, 11, 55-75.

42. Murry, H.H. (2000). "Gold: Progress in Chemistry, Biochemistry and Technology Edited by Hubert Schmidbaur (Technical University of Munich, Germany)." *Journal of the American Chemical Society*, 122, 4534-4534.
43. Tolaymat, T. M.; El Badawy, A. M.; Genaidy, A.; Schenekel, K.G.; Luxton, T.P.; Suidan, M. 2010. "An evidence-based environmental perspective of manufactured silver nanoparticle in syntheses and applications: A systematic review and critical appraisal of peer-reviewed scientific papers." *Science of The Total Environment*, 408, 999-1006.
44. Khaydarov, R.R.; Khaydarov, R.A.; Evrhafova, S.; Wagner, S.; CHO, S. Y. (2011). "Environmental and Human Health Issues of Silver Nanoparticles Applications." *Environmental Security and Ecoterrorism*. In: ALPAS, H., BERKOWICZ, S. M. M. & ERMAKOVA, I. (eds.). Springer Netherland.
45. Lewis, L.N. (1993). "Chemical catalysis by colloids and clusters." *Chemical Reviews*, 93, 2693-2730.
46. Solo 'E , A.; Pote Hina, T.; Cherno A, I.; Basin, B. (2007). "Track membrane with immobilized colloid silver particles." *Russian Journal of Applied Chemistry*, 80, 438-442.
47. Niemeyer, C. M. (2001). "Nanoparticles, Proteins, and Nucleic Acids: Biotechnology Meets Materials Science." *Angewandte Chemie International Edition*, 40, 4128-4158.
48. David, D.; Evanoff, Jr.; George, C. (2005). "Synthesis and optical properties of silver nanoparticles and arrays." *Chem. Phys. Chem.* 6, 1221-1231.
49 Kazakevich, P.V.; Voronov, V.V.; Simakin, A.V.; Shafeev, G.A. (2004). "Production of copper and brass nano particles upon laser ablation in liquids." *Quantum Electron*, 34 (2004), 951-956.
50. Hunsberger, Lynn (1999). "UV-Vis Absorption Spectroscopy General Chemistry Laboratory Manual"
51. Shelby, Hatch; George, Schatz (2008). "Synthesis and analysis of Silver/Gold nanoparticles." Northwestern University, Evanston, IL 60208.
52. http://www.nanocomposix.com/services/aerosol-facility
53 Schasfoort, R.B.M.; Tudos, A.J. (2008). "Handbook of Surface Plasmon Resonance." *The Royal Society of Chemistry Publishing*, Cambridge, UK.
54- Pockrand, I.; Swalen, J. D.; Gordon II, J. G.; Philpott, M. R. (1978). "Surface Plasmon spectroscopy of organic monolayer assemblies." *Surface Science*, 74, 237-244.
55- Gordon II, J. G.; Ernst, S. (1980). "Surface plasmons as a probe of the electrochemical interface." *Surface Science*, 101, 499-506.
56. Li, W. R.; Xie, X. B.; Shi, Q. S.;, Zeng, H. Y.;, OuYang, Y. S.; Chen, Y. B. Appl Microbiol Biotechnol. 2010, 85(4), 1115-22
57- Ying, Jackie (2001). *Nanostructured Materials*. New York: Academic Press. ISBN 0-12-744451-3.
58- Hanley, C; Thurber, A; Hanna, C; Punnoose, A; Zhang, J; Wingett, DG; Thurber; Hanna; Punnoose; Zhang; Wingett (2009). "The Influences of Cell Type and ZnO Nanoparticle Size on Immune Cell Cytotoxicity and Cytokine Induction". *Nanoscale Res Lett* **4** (12): 1409–20.
59. Wigginton, N. S., Haus, K. L. & Hochella JR, M. F. (2007). "Aquatic environmental nanoparticles." *Journal of Environmental Monitoring*, 9, 1306-1316.
60. Domingos, R. F.; Baalousha, M. A.; Ju-nam, Y., Reid, M. M.; Tufenji, N.; Lead, J. R.; Leppard, G. G.; Wilkinson, K. J. (2009). "Characterizing Manufactured Nanoparticles in the Environment: Multimethod Determination of Particle Sizes." *Environmental Science & Technology*, 43, 7277-7284.
61. David, D.; Evanoff, Jr.; George, C. 2005. "Synthesis and optical properties of silver nanoparticles and arrays." *Chem. Phys. Chem.* 6, 1221-1231.
62. Henglein, A. (1993). "Physicochemical properties of small metal particles in solution: microelectrode reactions, chemisorptions, composite metal particles and the atom-to-metal transition" *J. Phys. Chem.*, 97 (1993), 5457–5471.
63. P. Mulvaney, Surface plasmon spectroscopy of nanosized metal particles, Langmuir, 12 (1996), 788-800.
64. Mamun Ur Rashid (2009). "Synthesis of Silver Nano Particles (Ag-NPs) and their uses for Quantitative Analysis of Vitamin C Tablets."
65. Solomon, Sally D., Mozghan Bahadory, Aravindan V. Jeyarajasingam, Susan A. Rutkowsky, and Charles Boritz (2012). "Synthesis and Study of Silver Nanoparticles." *Journal of Chemical Education* 84.2 : 322-25. ACS Publications. Web. 9 Mar. 2012
66. Yusuf, Nur (2011). "Gold Nanoparticles: Synthesis, Characterisation and their Effect on Pseudomonas Flourescense." Thesis, University of Birmingham of Birmingham, (p-93)
67. Bonsak, Jack (2010). "Chemical Synthesis of Silver Nanoparticles for Light Trapping Applications in Silicon Solar Cells." Thesis. University of Oslo, 2010. Web. 9 Mar. 2012.

68. Mathew Wynn (2010). "The Synthesis and Characterization of Gold and Silver Nanoparticles in Formal and Informal Settings." Materials Engineering Department California Polytechnic State University, San Luis Obispo.
69. Das, R., S. S. Nath, D. Chakdar, G. Gope, and R. Bhattacharjee (2012). "Preparation of Silver Nanoparticles and Their Characterization." . <http://www.azonano.com/article.aspx?ArticleID=2318>.

436278 Immobilization and Catalytic Activity of Lipase on ModifiEd MWCNT for Oily Wastewater Treatment

Abstract:

We report here on the results of a study of immobilization of Lipase enzyme from *Candida Rugosa* on Multiwall Carbon Nanotubes (MWCNT) locally produced in a Chemical Vapor Deposition (CVD) reactor. Covalent bonds was formed between the enzyme and MWCNTs using organic cross linkers (NHS and EDC). FTIR results confirmed the covalent binding of the amide functional group of the lipase molecules onto the surface of MWCNT. The produced carbon nanotubes proved to be a good matrix for the lipase loading and immobilization. TGA results and calculations confirmed that 30% wt% of the enzyme was successfully attached to the functionalized MWCNTs. The nanocomposite material revealed high thermal stability and good hydrophobic characteristics which is considered an important property for commercial applications. Findings revealed higher than generally reported biological activity of the protein was successfully retained. Results confirmed that Immobilized lipase on MWCNTs present an excellent opportunity for treatment of oily wastewater. Teste with synthetic water/oil emulsion resulted with oil degradation reached up to 98%.

QUALITATIVE SYNTHENSISATION OF NANOPARTICLES USING FTIR SPECTROSCOPIC ANALYSIS

Engr. Dr. M. N. Idris* and **K. M. Gwary**
Energy & Process Refining Technology Group
Department of Chemical Engineering
University of Maiduguri
Borno State, Nigeria
Tel: +234 (0) 705 877 1096
Email: idrismn@unimaid.edu.ng, muhinu@hotmail.co.uk

ABSTRACT

In this work, FTIR spectrometry was used to analyse the spectra configurations on the synthesised silver nanoparticles were synthesized using two different reducing agents (1) sodium borohydride ($NaBH_4$), at two temperature conditions of 27^{0}C and 4^{0}C (room temperature and ice cold temperature), and (2) trisodium citrate (3NaCi) at 100^{0}C (boiling temperature) by chemical reduction method.

The resulting nano-sized colloids have been characterized by FTIR spectroscopy. A clear and yellow colouration was obtained from all the synthesised samples, which indicates the presence of formed nanoparticles with strong absorption peaks obtained at wave length range of 1370 - 1380 cm^{-1} and 840 - 841 cm^{-1} proving the presence of silver nanoparticles based on the functional group. The elements found to be present at such peak were the nitro and N-O functional groups.

Keywords: *Synthesis, FT-IR, Nanoparticle, reducing agent and functional group*

***Correspondence about this article**: Please for all correspondence on this article, kindly contact the principal author as stated above. Thank you.*

HYDROGEN PEROXIDE STABILITY IN SILICA HYDROGELS

Fulya Sudur, Nancy Goodyear, Nese Orbey, University of Massachusetts Lowell, Lowell, MA, USA

Abstract

Hydrogen peroxide (H_2O_2) is an environmentally friendly oxidant whose current applications are limited due to its instability. Silica hydrogels obtained by sol-gel process have the potential of increasing its stability due to strong hydrogen bonding between the H_2O_2 molecules and the silanol groups (Si-OH) at the silica surface. The pH of the sol determines the structure of the hydrogel; in previous studies it is controlled by adding an acidic or alkaline solution. In the present study, the pH of the sol is adjusted by varying the sodium content of sodium silicate. The effect of sodium on the kinetics of sol-gel formation and the structure of the resulting hydrogels are studied. The surface area of the hydrogel/xerogel determines the availability of sites for H_2O_2 bonding in the presence of sodium ions. Therefore, the decomposition rate of H_2O_2 is studied at different sodium concentrations to understand its stability at room temperature. Samples are characterized by measuring the surface area, pore volume and average pore size using Brauner-Emmet-Teller (BET) analysis and scanning electron microscopy (SEM). The antimicrobial activity of the hydrogels containing 10 and 20 wt% H_2O_2 was evaluated against *Escherichia coli* and *Staphylococcus aureus* and the results demonstrated that hydrogels exert a potent microbicidal efficacy.

Introduction

Hydrogen peroxide (H_2O_2) is a green oxidant and an antimicrobial agent. It has broad range of industrial applications, such as pulp and textile bleaching, teeth whitening, skin disinfection, and water treatment.[1-4] The use of H_2O_2 in current applications is limited due to its instability according to the following reaction:

$$H_2O_{2(aq)} \rightarrow H_2O_{(l)} + \tfrac{1}{2}O_{2(g)} \quad (1)$$

The stability of H_2O_2 can be increased by entrapping it in silica xerogels because hydrogen peroxide forms strong hydrogen bonds with the silica gel surface.[5,6] The xerogels are formed via sol-gel chemistry by using an aqueous silicate precursor, sodium silicate.[7] The first step of synthesizing silica gels is to form a colloidal suspension, also defined as sol, by removing sodium ions from the sodium silicate with a cation exchange resin. The sodium content of the sol can be adjusted using an ion exchange process,[8-10] as shown below,

$$Na^+ + RzSO_3^- H^+ \rightarrow H^+ + (RzSO_3)Na \quad (2)$$

where Rz is the styrene-divinylbenzene copolymer and SO_3H is the sulfonic acid group. Ion exchange occurs easily because the ionic bond strength of a sodium ion is stronger than that of a hydrogen ion.[10]

Further bonding between the silanol groups leads to the formation of a three-dimensional interconnected network resulting in a "wet gel" or "hydrogel". Xerogels are formed simply by drying the hydrogels. The pH of the sol affects the inorganic polymerization, and, hence influences the properties of the silica network.[11]

Silica xerogels containing H_2O_2 have been reported in the literature. Żegliński el al.[5] synthesized xerogels containing 3.8 to 68 wt% H_2O_2. The stability of 3.8 wt% H_2O_2 in xerogel was 71% after 63 days of storage at room temperature and as high as 94% after storage at 3 °C. Bednardz et al.[6] used H_2O_2 containing silica xerogels as oxidants in organic synthesis. The xerogels were stored at

5 °C and half of the initial H_2O_2 loading of 8 wt% was retained. Room temperature stability of H_2O_2 was not reported.

In the present work, the effect of sodium on the kinetics of sol-gel formation in the presence of H_2O_2, its consequent effect on its stability and the structure of the resulting hydrogels are studied. H_2O_2 stability in silica hydrogels at room temperature was monitored at periodic intervals by potassium permanganate ($KMnO_4$) titration. Hydrogels were characterized by scanning electron microscopy (SEM) and Brunauer-Emmett-Teller (BET) analysis. The antimicrobial activity of the hydrogels at various H_2O_2 concentrations was examined.

Experimental Procedure

Materials

Sodium silicate solution (reagent grade, 10.6% Na_2O, 26.5% SiO_2); Amberlite IR120 (hydrogen form); hydrogen peroxide (contains inhibitor, 30% wt in H_2O, ACS reagent); orthophosphoric acid (85% wt in H_2O, 99.99% trace metals basis); potassium permanganate (ACS reagent >99%) and sulfuric acid (99.999%). All chemicals were purchased from Sigma Aldrich Chemical Co. and used without further purification.

Hydrogel Synthesis

Details of the hydrogel synthesis procedure are described elsewhere.[12] Briefly, as received sodium silicate solution was diluted with deionized water and combined with 30 wt% H_2O_2 solution and 7 μL of phosphoric acid (H_3PO_4). The sodium content of the sodium silicate solution was controlled by adding different amounts of ion exchange resin (Amberlite IR 120, hydrogen form) to the sol. The mixture was stirred with an overhead mixer, the resin was allowed to settle, and the sol was transferred to a glass Petri dish (D = 100 mm, h = 15 mm) and placed in a BlueM gravity oven at 40 °C until gelation was observed.

The effect of sodium concentration on the stability of H_2O_2 was studied keeping the initial H_2O_2 concentration constant at 20 wt% and changing the sodium concentration between 0.027 and 0.557 wt%. The effect of initial H_2O_2 concentration on the stability of H_2O_2 was also studied at concentrations between 10 and 20 wt% while keeping the sodium concentration constant at 0.027 wt%.

Characterization

The sodium content of the sol before gelation was measured using a flame atomic absorption spectrometer (240 AA, Agilent Technologies). The H_2O_2 content of the hydrogels was determined by $KMnO_4$ titration immediately after gelation and at periodic intervals. Three repeated titrations were performed on each sample to check reproducibility.

The morphology of the hydrogels was studied using a JEOL JSM 7401F scanning electron microscope. The surface area, pore size, and pore size distribution were obtained through BET analyses by first converting into aerogels using a critical-point dryer (Tousimis SAMDRI-795) to preserve the solid network and retain the pore structure.

Antimicrobial activity was determined by agar well diffusion. Briefly, *Escherichia coli* (ATCC 29214, American Type Culture Collection, Manassas, VA) and *Staphylococcus aureus* (ATCC 6538) were grown overnight at 37°C in Tryptic Soy Broth (Becton Dickinson Co, Sparks, MD). The overnight growth was standardized spectrophotometrically to ~1.5 x 10^8 colony forming units/mL by dilution in sterile phosphate buffered saline (PBS). The standardized bacterial solution was inoculated to duplicate Tryptic Soy Agar plates (Becton Dickinson Co, Sparks, MD) to achieve an even lawn of growth over the entire plate. A 6 mm well cutter was used to punch a single well in the middle of each inoculated plate. The wells were filled to the top with hydrogels containing 10 wt% and 20 wt% H_2O_2 at same sodium content (0.027 wt%). The control sample contained equivalent amount of silica hydrogel without H_2O_2. After overnight incubation at 37°C, the diameters of the zones of inhibition were measured in mm and the average and standard deviation of the duplicates calculated.

Results

Sodium Content

In this study, the amount of residual sodium in the silicic acid determined the pH of the sol whereas in the previous studies the pH was controlled by adding acidic or alkaline solution.[13]

Figure 1 shows a decrease in the pH of the sol at lower sodium contents because sodium ions were replaced with hydrogen ions as a result of ion exchange. A sharp decline in pH was observed at sodium contents below 0.1 wt%. We believe this is due to the high buffer capacity of the sol above 0.1 wt%.[14] In other words, the low pH indicates the low sodium content of the sol.

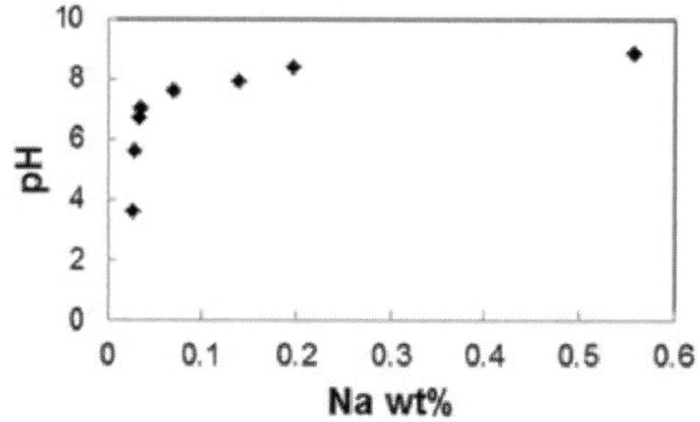

Figure 1. Relationship between the pH and Na content of the sol (initial H_2O_2 concentration 20 wt%)

Relationship between gel time and pH

In this study, the gel time was determined based on the procedure reported in the literature. The moment when the gel broke away from the wall of the Petri dish, instead of flowing as a liquid when the dish was tilted, was considered as the gelation point.[15, 16]

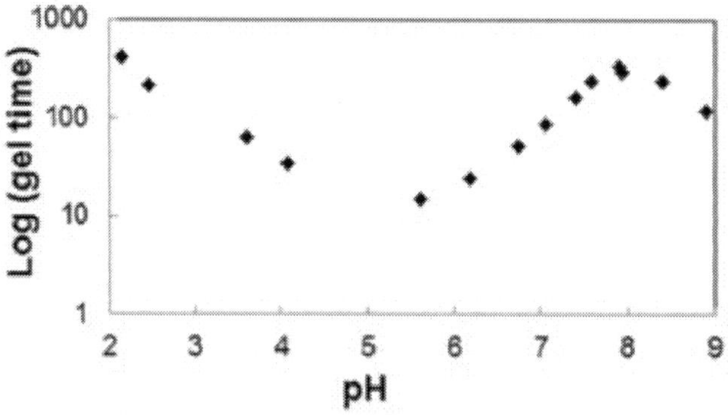

Figure 2. Relationship between gel time and pH (initial H_2O_2 concentration 20 wt%)

The effect of pH on the gel time of silicic acid-water-H_2O_2 was studied and the results are plotted in Figure 2. Low concentration of hydroxyl ions at pH range 2-4 resulted in a slow gel formation. As pH was increased from 4 to 6, the net charge of the sol was zero resulting in a rapid gelation. After the sol passed through the isoelectric point (pH 6–8), the concentration of hydroxyl ions increased and longer gel times were obtained. The presence of sodium ions affected the gelation above a pH of 8 resulting in a decrease in gel time.[17]

When compared to the ones reported in the literature, our results shifted slightly towards the acidic side due to the presence of H_2O_2 in the sol.[18-20]

The effect of H_2O_2 concentration on the gel time was also studied and found that the gel time decreased as the H_2O_2 concentration increased.

Stability of H_2O_2 at room temperature

Decomposition of H_2O_2 followed a first-order rate law during short term storage (10 days):

$$-\frac{d[H_2O_2]}{dt} = k_{H_2O_2}[H_2O_2] \quad (3)$$

$$\ln\left(\frac{[H_2O_2]}{[H_2O_2]_0}\right) = -k_{H_2O_2}t \quad (4)$$

where $k_{H_2O_2}$ is the observed first-order rate constant, and $[H_2O_2]$ and $[H_2O_2]_0$ are the H_2O_2 concentrations in the hydrogels at any time t and time zero, respectively.

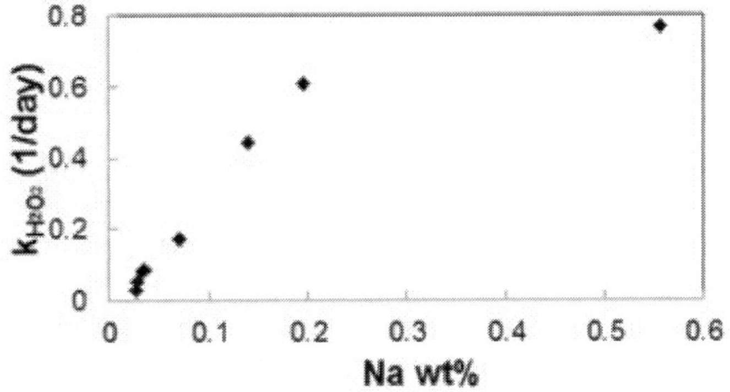

Figure 3. Dependence of the decomposition rate constant on the sodium content of the hydrogels (at 20 wt% H_2O_2)

Figure 3 shows that as the sodium content decreased, the H_2O_2 decomposition rate also decreased. The decomposition rate constant was the same for the hydrogels having different H_2O_2 concentrations at a given sodium content for short term storage.

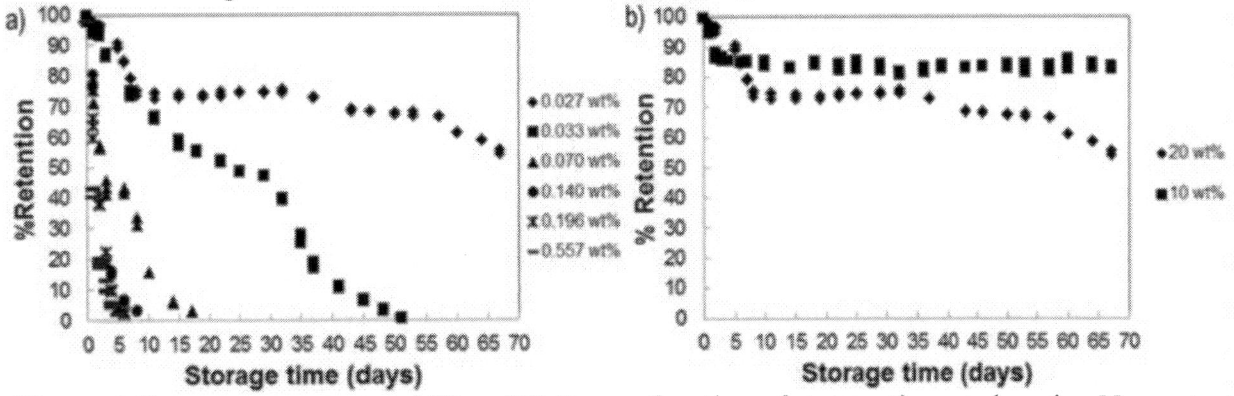

Figure 4. Room temperature stability of H_2O_2 as a function of storage time at changing Na content (initial H_2O_2 concentration 20 wt%) and at changing H_2O_2 concentration (Na content 0.027 wt%)

The stability of entrapped H_2O_2 in hydrogels containing different sodium and H_2O_2 was monitored for 70 days at room temperature. Figure 4a shows that the highest sodium content hydrogel resulted in the fastest H_2O_2 decomposition (5 days) whereas the lowest sodium content hydrogel retained more than 50% of H_2O_2 after 70 days of storage. On the other hand, at low sodium concentration when H_2O_2 concentration decreased to 10 wt%, the retention values were significantly enhanced (up to 89%) as shown in Figure 4b. The optimized sodium content and gelation conditions (time and temperature) resulted in higher stability of H_2O_2.

Morphology and structure of hydrogels

BET and SEM analyses revealed that sodium affected the structure of hydrogels. The samples were prepared by forming aerogels using supercritical CO_2. The effect of sodium content on the surface area, pore volume, and average pore diameter obtained from BET analysis are reported in Figure 5.

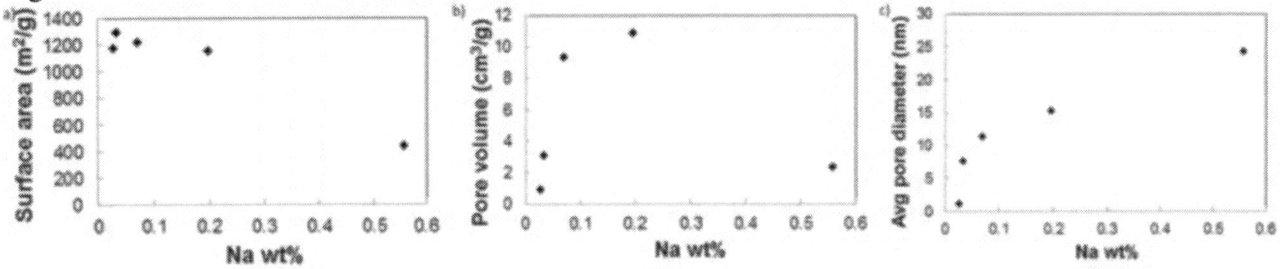

Figure 5. Structural properties of silica hydrogels with changing sodium content (initial H_2O_2 concentration 20 wt%)

Figure 5a shows the surface area measurements where the low sodium concentrations did not have an appreciable effect. However, at higher sodium contents a sharp decrease in the surface area was observed. We believe that silica dissolves and reprecipitates at high sodium content (and at high pH) caused a reduction in the surface area and formation of pores with larger diameter.[21, 22]

We observed an initial increase and then a decrease in the pore volume as a function of sodium content (Figure 5b). Pore volume increased with the sodium content resulting in a decrease in H_2O_2 stability. H_2O_2 stability was also low when the pore volume was low at high sodium contents. These results indicate that sodium has a more significant effect on stability than the pore volume.

Figure 5c shows that sodium content and the pore diameter have a linear relationship. The dissolution and reprecipitation of silica along with the interparticle condensation as a result of high sodium content resulted in larger pores. Formation of smaller sized pores was observed due to the low dissolution rate of silica at low sodium concentrations.

The effect of the H_2O_2/water ratio on the surface area, pore volume and average pore diameter was much less pronounced than the effect of sodium content.

Figure 6. SEM images of hydrogels containing different sodium contents: a) 0.557 wt%, b) 0.196 wt%, c) 0.027 wt% (initial H_2O_2 concentration 20 wt%)

The morphology of the hydrogels with different sodium content is shown in Figure 6. The SEM images indicate that silica network formation was affected by the sodium presence due to changes in hydrogen bonding capacity and strength. At low sodium contents, more compact and denser gel structure was obtained.

Antimicrobial Activity

The test results obtained for the efficacy of our hydrogels containing 0, 10 and 20 wt% H_2O_2 with *E. coli* and *S. aureus* are presented in Table 1 where the average diameters of zones of inhibition (ZOI) are listed. Figure 7 shows the results obtained for *S. aureus* only.

Table 1. Average diameters of zones of inhibition (ZOI) for each hydrogel tested

	Average ZOI Diameter in mm (SD)	
Silica hydrogels	***E. coli***	***S. aureus***
with water	16 (±1.4)	19 (±1.4)
with 10 wt% H_2O_2	43 (±1.4)	51.5 (±2.1)
with 20 wt% H_2O_2	49 (±1.4)	59.5 (±2.1)

The results indicate that bacterial growth was inhibited for both *E. coli* and *S. aureus* and that the ZOI diameter increases with increasing H_2O_2 content in the hydrogel. Since the structure of the hydrogels containing 10 and 20 wt% H_2O_2 was very similar to each other (based on the discussion in the previous section), we believe that the difference obtained from the results is due to the variation in H_2O_2 concentrations. The hydrogels containing water only also exhibit antibacterial properties. We think that the presence of the gel by itself inhibits the bacterial growth.

The mean ZOI values for *S. aureus* were slightly higher than those for *E. coli.* The suppression of the growth of *S.aureus* was attributed to the structural differences between the two bacteria.[23]

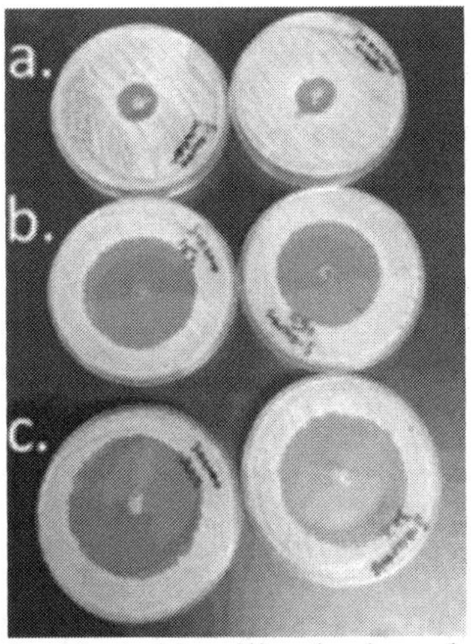

Figure 7. Growth of *S.aureus* when incubated in silica hydrogel a) with water, b) with 10 wt% H_2O_2, c) with 20 wt% H_2O_2

Conclusion

H_2O_2 was successfully stabilized in a silica hydrogel matrix. Our results indicate that the sodium content of the sol has a significant effect on the structure of hydrogels. Mesoporous (3-24 nm) hydrogels with a large surface area (1000-1400 m^2/g) were obtained. Sodium content also affected the retention and stability of H_2O_2 and its decomposition followed a first-order reaction. Optimizing the sodium content and gelation protocol (time and temperature) resulted in higher retention and better stability of H_2O_2 at room temperature. The hydrogels showed antimicrobial activity.

References

1. Colodette, J. L.; Rothenberg, S.; Dence, C. W. Factors Effecting Hydrogen Peroxide Stability in the Brightening of Mechanical and Chemimechanical Pulps. *J. Pulp. Pap. Sci.* 1989, 15, 3-10.
2. Sulieman, M.; Addy, M.; MacDonald, E.; Rees, J. S. The Effect of Hydrogen Peroxide Concentration on the Outcome of Tooth Whitening: An In Vitro Study. *J. Dent.* 2004, 32, 295-299.
3. Kim, J.C.; Lee, K. U.; Shin, W. C.; Lee, H. Y.; Kim, J. D.; Kim, Y. C.; Tae, G.; Lee, K. Y.; Lee, S. J.; Kim, J. D. Monoolein Cubic Phases Containing Hydrogen Peroxide. *Colloids Surf., B* 2004, 36, 161-166.
4. Patil, A. R.; Kausley, B. S.; Balkunde, L. P.; Malhotra, P. C. Comparative Study of Disinfectants for Use in Low Cost Gravity Driven Household Water Purifiers. *J. Water Health* 2013, 11.3, 443-456.
5. Żegliński, J.; Cabaj, A.; Strankowski, M.; Czerniak, J.; Haponiuk, J.T. Silica Xerogel-Hydrogen Peroxide Composites: Their Morphology, Stability, and Antimicrobial Activity. *Colloids Surf., B* 2007, 54, 165-172.
6. Bednardz, S.; Rys, B.; Bodgal, D. Application of Hydrogen Peroxide Encapsulated in Silica Xerogels to Oxidation Reactions. *Molecules* 2012, 17, 8068-8078.
7. Baccile, N.; Babonneau, F.; Thomas, B.; Coradin, T. Introducing Ecodesign in Silica Sol-Gel Materials. *J. Mater. Chem.* 2009, 19, 8537-8559.
8. Bergna, H. E.; Robertw, W. O. *Colloidal Silica: Fundamentals and Applications*; CRC Press: Florida, 2005.
9. Van Dokkum, H.P.; Hulskotte, H. J.; Kramer, K. J. M; Wilmot, J. Emission, Fate and Effects of Soluble Silicates (Waterglass) in The Aquatic Environment. *Environ. Sci. Technol.* 2004, 38, 515-521.
10. Kim, T.; Hwang, S.; Hyun, S. Development of a Continuous Manufacturing Process for Silica Sols via the Ion Exchange of a Waterglass. *Ind. Eng. Chem. Res.* 2008, 47, 6941-6948.
11. Quarch, K.; Kind, M. Inorganic Precipitated Silica Gel. Part 1: Gelation Kinetics and Gel Properties. *Chem. Eng. Technol.* 2010, 33, 1034–1039.
12. Sudur, F; Pleskowicz, B.; Orbey, N. Hydrogen Peroxide Stability in Silica Hydrogels. *Ind. Eng. Chem. Res.* 2015, 54(6), 1930-1940.
13. Iler, R. K. *The Colloidal Chemistry of Silica and Silicates*; Cornell University Press: New York, 1955.
14. Hamouda, A. A.; Amiri, H. A. A. Factors Affecting Alkaline Sodium Silicate Gelation for In-Depth Reservoir Profile Modification. *Energies* 2014, 7, 568-590.

15. Merrill, R. C.; Spencer, R. W. The Gelation of Sodium Silicate. Effect of Sulfuric Acid, Ammonium Sulfate, and Sodium Aluminate. *J. Phys. Chem.* 1950, 54, 806-812.
16. Munoz-Aguado, M. J.; Gregorkiewitz, M. Sol-Gel Synthesis of Microporous Amorphous Silica from Purely Inorganic Precursors. *J. Colloid Interface Sci.* 1997, 185, 459-465.
17. Berrier, E.; Courtheoux, L.; Bouazaoui, M.; Capoen, B.; Turrell, S. Correlation between Gelation Time, Structure and Texture of Low Doped Silica Gels. *Phys. Chem. Chem. Phys.* 2010, 12, 14477-14484.
18. Iler, R. K. *The Colloidal Chemistry of Silica and Silicates*; Cornell University Press: New York, 1955.
19. An-Pang, T. A Theory of Polymerization of Silicic Acid. *Sci. Sin.* 1963, 12, 1311-1320.
20. Brinker, J. C.; Scherer, G. W. *Sol-Gel Science: The Physics and Chemistry of Sol-Gel Processing;* Academic: San Diego, 1990.
21. Baxter, S.; Bryant, K.C. Silica Sols. Part III. Accelerated Gelation and Particle Size. *J. Chem. Soc.* 1952, 74, 3024-3027.
22. J.; Deshpande, R.; Smith, D. M; Brinker, C. J.; Assink, R. A. Pore Structure Evolution in Silica Gel During Aging/Drying. IV. Varying Pore Fluid pH. *J Non-Cryst. Solids* 1994, 167, 295-306.
23. Shilpakala, S.R.; Prathiba, J.; Malathi, R. Susceptibilities of *Escherichia coli* and *Staphylococcus aureus* to *Aloe barbadensis*. *Eur. Rev. Med. Pharmacol. Sci* 2009, 13, 461-464.

424076 Prediction of Nanoparticles-Cell Association Based on Corona Proteins and Physicochemical Properties

Abstract:

The association of nanoparticles (NPs) with cells in biological fluids is impacted by the adsorption of proteins onto the NP surface. The multitude of proteins in the biological fluid form a protein corona which can then affect cellular bioactivity. In the present work, an extensive dataset of gold NPs protein corona was utilized to investigate the relationships between protein corona fingerprints (PCFs) and NP-cellular association. The dataset consisted of a library of 84 gold nanoparticles (NPs) of 15, 30, or 60 nm cores with cationic or anionic surface ligands, and data regarding the abundance of 129 specific corona proteins and 19 NO physicochemical properties. The specific protesins of relavance to NP-cellular association were identified via quantitative structure-activity relationships (QSARs) that were developed based on both non-linear support vector regression (SVR) and linear models. Feature selection was accomplished via sequential forward floating selection. From the initial set of 148 descriptors, the serum proteins *APOB*, *A1AT*, *ANT3*, and *PLMN* in addition to the NP zeta potential were most significant for correlating cellular association of gold nanoparticles. The non-linear model was built to a reasonable level of accuracy with only 6 serum proteins and NP zeta potential with a performance accuracy of R^2=0.895. In contrast, the linear model required 11 CFs and demonstrate performance of R^2=0.850. Model robustness was validated based on Y-randomization demonstrating that the QSARs were not "chance" correlations. The present approach demonstrates that exploration of NP-cell association data, via QSAR analysis, can be particularly useful to elucidate the role of NP corona proteins and NP physicochemical properties in affecting NP-cellular. Such information is critical for supporting the design and interpretation of toxicity studies and for the design of NPs for biomedical applications.

428378 Characterization of Detergent Residues on Surfaces

Abstract:

Characterization of detergent residues on surfaces

Sadi Gurses, Alper Uzun, Can Erkey, Seda Kizilel

Department of Chemical and Biological Engineering, Koc University, Istanbul 34450, Turkey

The problem of filmy appearance on dishwasher-washed dishes has been receiving significant attention with increased health awareness of consumers. There is a need for characterization of dishware surfaces following a complete wash cycle. Here, we focused on identifying the indicator chemical components to detect detergent residues on the dishware surfaces and their toxicity on live cells. Spectroscopic and diffraction measurements have been carried out to determine structural information about detergent components, Fourier transform infrared spectroscopy (FT-IR) and high pressure liquid chromatography (HPLC) have been used to analyze rinsing water. Surfaces of the dishware were analyzed with X-ray photoelectron spectroscopy (XPS) and inductively coupled plasma mass spectroscopy (ICP-MS). Surface analysis results showed decomposition of penta sodium tripolyphosphate and sodium carbonate, known as builders in detergent chemistry, on surfaces upon dishwashing process. We also performed toxicity assays, where the influence of these builders on human fibroblast cell viability was investigated. Cell viability results showed decreasing trend in the number of cells cultured with increasing concentrations and exposure time of penta sodium tripolyphosphate in the medium. This study is significant for characterization of unknown detergent residues on surfaces and their possible toxicity on live cells.

421609 Assessment on the Treatment of Carbon Soot for Activated Carbon Applications: Oxidative Stress and Apoptosis Induced in Human Cell Lines

Abstract:

In Singapore, approximately 30 ton/day of carbon based solid waste is produced from petroleum chemical process. Solid waste like carbon soot is generally disposed to landfill sites or re-utilized as secondary raw materials. Carbon soot has gained much attention in its re-utilization due to its high content of carbon for possible processing as absorbent materials. However, carbon soot also contains hazardous components such as heavy metals arising from either sources or petroleum chemical process. These components may be toxic to human health and environment, and pose many restrictions in its re-utilization due to the potential toxicity. Therefore, it is important to examine the possible toxicity effects of carbon soot and its mechanism on human health. In this study, we performed chemical and microbial leaching on carbon soot to remove hazardous components before further processing it as activated carbon. Subsequently, three human cell lines i.e. HepG2, MRC-5 and MDA-MB-231 were used to investigate the cytotoxicity, oxidative stress and apoptosis induction by carbon soot. Carbon soot induced dose-dependent and time-dependent cytotoxicity in HepG2, MRC-5 and MDA-MB-231 cells as demonstrated by MTS and LDH assays. Meanwhile, the flow cytometry assay and fluorescence microscope results indicated that the carbon soot induced apoptosis of HepG2, MRC-5 and MDA-MB-231 cells increased with the increasing concentration of carbon soot. Carbon soot was also found to induce oxidative stress in HepG2, MRC-5 and MDA-MB-231 cells as demonstrated by generation of reactive oxygen species (ROS) and the alteration of antioxidant superoxide dismutase (SOD) and glutathione (GSH) compared to the unexposed control cells. Further, co-treatment with N-(mercaptopropionyl)-glycine (N-MPG), a type of ROS scavenger formulations, could inhibit cells apoptosis induced by carbon soot and increase the viability of cells. These results may provide an explanation for the potential mechanism in HepG2, MRC-5 and MDA-MB-231 cytotoxicity under the effect of carbon soot. We hope that these findings could help to understand, in a microscopic scale, the threat and risk of these wastes to human.

427903 Structure-Toxicity Correlations for Ni/SiO2 Complex Engineered Nanomaterials Using High-Throughput Zebrafish Assays

Abstract:

Functional nanomaterials are developing at an ever accelerating pace and are finding application in a wide range of industrial and consumer products despite evidence that nanomaterials can show elevated toxicity. Furthermore, while bare nanoparticles (NPs) are rarely used in industrial application due to sintering and deactivation effects, embedded NPs (so-called complex engineered nanomaterials, CEN) have found little attention in nanotoxicological studies to-date. This motivates a need for the development of rapid, yet sensitive, high throughput *in vivo* nanotoxicity screening assays.

We are investigating three differently structured Ni/SiO_2 nanomaterials as model CEN with the aim to evaluate different nanotoxicity assays and develop structure-toxicity correlations for CEN that will allow derivation of predictive toxicity models. The CEN are comprised of Ni NPs embedded in or on (non-toxic) porous silica NPs, based on the hypothesis that embedding nickel nanoparticles in silica could reduce or entirely mitigate nanotoxicity while still providing accessibility and hence maintaining functionality of the embedded NP. The three CEN structures are nickel NPs (a) embedded in hollow porous silica shells (hNi@SiO_2), (b) encapsulated in non-hollow, porous silica NPs (nhNi@SiO_2), and (c) deposited on the silica nanoparticle's external surface (Ni-SiO_2). All CENs were thoroughly characterized via TEM, XRD, BET, as well as for dissolution, aggregation, and settling properties, followed by multiple 5-day zebrafish developmental toxicity assays including survival, malformation, hatching, and motility. Zebrafish (*Danio rerio*) were used as a well-suited toxicity model due to prolific breeding and rapid development time. The toxicity results were then correlated with the CEN properties to determine structure-toxicity correlations.

All three nanomaterials mitigated toxicity compared to the respective Ni^{2+} dose and showed high zebrafish survival. Zebrafish motility, which probes persistent neurotoxicity, emerged as a more sensitive assays and revealed lowest toxicity for nhNi@SiO_2, explained by lower nickel dissolution compared to the other two nanomaterials. In contrast, this material showed the highest toxicity of the three CENs in hatching assays, which can be traced back to resistance to aggregation which facilitates uptake through the chorion. Overall, the toxicity results are consistent with a 'Trojan horse' mechanism and point towards complexity and pitfalls when establishing structure-toxicity correlations and the necessity of combining multiple assays to fully assess nanomaterials toxicity.

430535 Size-Dependent Cellular Toxicity of Gold Nanoparticles on Human Embryonic Stem Cells

Abstract:

Increased demands for toxicological evaluations of engineered nanomaterials coincide with the accelerated pace of clinical translation and commercialization of nanomaterials. There is an urgent need to develop predictive and validated toxicological assessment methods for nanomaterials due to their unique size range, surface chemistry, and interactions with biological systems. Most cell-based nanotoxicity assessments are limited to long-lived cell lines or cancer cells. Human embryonic stem cells (hESCs) and their subsequent neural precursor differentiations provide a toolkit to study the effect of exposure to nanoparticles on neural specification. We employ gold nanoparticles (AuNPs) as model nanoparticles due to their versatile surface chemistry, ease of imaging, and tunability for transport across the blood-brain barrier. AuNPs are considered by most as nontoxic, which is consistent with the inertness of bulk gold, but some have found them toxic. This talk describes a toxicological study of AuNPs of different core sizes (1.5 nm, 4.0 nm, and 14 nm) as well as dendrimers (0.5 nm and 1.5 nm) using the hESCs. The various nanoparticles were coated with bovine serum albumin prior to their introduction to the cell culture medium. The nanoparticles were characterized by UV-vis spectroscopy, TEM, dynamic light scattering, and AFM. We found that colonies exposed to 1.5 nm AuNPs exhibited loss of cohesiveness, rounding up, and detachment suggesting ongoing cell death. The hESCs exposed to 1.5nm AuNPs did not form embryoid bodies but rather rapidly disintegrated into single cells within 48 hours of treatment. Cell death caused by 1.5nm AuNPs also occurred in the treatment hESC-derived neural precursor/progenitor cells. None of the other nanoparticles exhibited such toxic effect on the hSECs. We also found that 4.0 nm AuNPs caused a dramatic hypomethylation of the hESC DNA in only 24 hours. This work contributes a new toolkit for predicting neurotoxicity of nanoparticles that will ultimately impact consumers, patients, and manufacturing workers of products containing nanoparticles.

418163 Micrornas As Biomarkers for Cytotoxicity Evaluation of Multi-Wall Carbon Nanotube

Abstract:

Multi-wall carbon nanotubes (MWCNT) are one of the most promising engineered nanomaterials because of their excellent physical and chemical characteristics. However, the applications of MWCNT also raise health concerns. The small size and light weight of MWCNT allow them to enter the working environment as particulate matter of respirable sizes. Many studies have already reported that inhaled CNTs showed significant cytotoxicity both in vitro and vivo, and the toxicity potential will increase with decreasing particle sizes.

MicroRNAs, a family of short endogenous noncoding RNAs, harbor critical functions in many fundamental biological processes, such as proliferation and apoptosis. Intracellular and circulating microRNAs have been implicated as promising biomarkers in the diagnosis of lung cancer and pulmonary diseases. Here, we seek to identify both intracellular and circulating microRNAs as biomarkers for MWCNT induced pulmonary toxicity.

Total 4 types of MWCNT with different lengths (0.2-2μm and 10-50μm) and surface modifications (with and without carboxyl modification) were selected and their cytotoxicity was evaluated in human bronchial epithelial cells (BEAS-2B). The cytotoxic effects of 4 types of MWCNT were first characterized by the cell growth measurement and the reactive oxygen species (ROS) assay. BEAS-2B cells were treated with MWCNT at concentrations from 0 to 50μg/mL. At 48 h post treatment, all 4 types of MWCNT inhibited the growth of BEAS-2B cells and generated significant ROS stress. The MWCNT toxicity was dose and tube type dependent. The shorter length and carboxyl modification enhanced the MWCNT toxicity because of better dispersion in aqueous solutions and higher cellular uptake.

Next, a microRNA toxicology panel was used to identify the dysregulated microRNAs post MWCNT exposure. Profiling of 87 miRNAs related with toxicology showed that the expression of total 7 miRNAs was changed at least two folds compared with untreated controls. Among 87 miRNAs, miRNA-29b, miRNA-21, and miRNA-486 were selected for further investigation because they are closely related with lung diseases including pulmonary fibrosis and lung cancer. At 48 h post treatment, the expression of intracellular miRNA-29b was up-regulated up to ~10 folds, the expression of intracellular miR-486 was down-regulated up to ~55% and the expression of intracellular miR-21 remained unchanged compared with untreated controls. However, the expression of circulating miRNA-29b was up-regulated up to ~4 folds, the expression of circulating miR-486 was up-regulated up to ~10 folds and the expression of circulating miR-21 was up-regulated up to ~50 folds compared with untreated controls. The expression of intracellular and circulating miRNAs was also dose and tube type dependent. Among 4 types of MWCNT, short MWCNT with carboxyl modification showed the most significant impact on the expression of both intracellular and circulating miRNAs. And the expression of intracellular and circulating miRNAs were able to distinguish the toxicity effects induced by different types of MWCNT.

Our results demonstrated that intracellular and circulating microRNAs are sensitive biomarkers for MWCNT induced pulmonary cytotoxicity, and they hold great potential in the early detection of adverse health effects associated with exposure to MWCNT.

AUTHOR INDEX

AIChE
120 Wall Street, FL 23
New York, NY 10005-4020

ISBN 978-1-5108-1848-4

9 781510 818484